Pergamon International Library
of Science, Technology, Engineering and Social Studies
The 1000-volume original paperback library in aid of education,
industrial training and the enjoyment of leisure
Publisher: Robert Maxwell, M.C.

The Radio Universe

THIRD EDITION

THE PERGAMON TEXTBOOK
INSPECTION COPY SERVICE

An inspection copy of any book published in the Pergamon International Library will gladly be sent to academic staff without obligation for their consideration for course adoption or recommendation. Copies may be retained for a period of 60 days from receipt and returned if not suitable. When a particular title is adopted or recommended for adoption for class use and the recommendation results in a sale of 12 or more copies, the inspection copy may be retained with our compliments. The Publishers will be pleased to receive suggestions for revised editions and new titles to be published in this important International Library.

Other Titles of Interest

J.P. BARBATO
*Atmospheres: A View of the Gaseous Envelopes
Surrounding Members of our Solar System*

R.S. KANDEL
Earth and Cosmos

A.T. LAWTON
A Window in the Sky

A.J. MEADOWS
Stellar Evolution

P.M. SOLOMONS & M.G. EDMUNDS
Giant Molecular Clouds in the Galaxy

A.J. WHYTE
The Planet Pluto

The Radio Universe

by

J.S. HEY

THIRD EDITION

PERGAMON PRESS

OXFORD · NEW YORK · TORONTO · SYDNEY · PARIS · FRANKFURT

U.K.	Pergamon Press Ltd., Headington Hill Hall, Oxford OX3 0BW, England
U.S.A.	Pergamon Press Inc., Maxwell House, Fairview Park, Elmsford, New York 10523, U.S.A.
CANADA	Pergamon Press Canada Ltd., Suite 104, 150 Consumers Road, Willowdale, Ontario M2J 1P9, Canada
AUSTRALIA	Pergamon Press (Aust.) Pty. Ltd., P.O. Box 544, Potts Point, N.S.W. 2011, Australia
FRANCE	Pergamon Press SARL, 24 rue des Ecoles, 75240 Paris, Cedex 05, France
FEDERAL REPLUBLIC OF GERMANY	Pergamon Press GmbH, Hammerweg 6, D-6242 Kronberg-Taunus, Federal Republic of Germany

---

First edition 1971

Second edition 1975

Third edition 1983

Library of Congress Cataloging in Publication Data

Hey, J.S.
The radio universe.
(Pergamon international library of science, technology, engineering, and social studies)
Includes index
1. Radio astronomy—Popular works. I. Title. II. Series.
QB477.H47 1983   522'.682   82-18982

British Library Cataloguing in Publication Data
Hey, J.S.
The radio universe.—3rd ed.
1. Radio astronomy I. Title 522'.682

ISBN 0–08–029152–X Hardcover
ISBN 0–08–029151–1 Flexicover

Printed in Great Britain by A. Wheaton & Co. Ltd., Exeter

Preface to the Revised Third Edition

SINCE the publication of the previous edition in 1975 rapid progress has been made in all branches of astronomy, and the combined results are bringing a new insight into astrophysical processes. There have been major advances in radio and optical astronomy as well as a breakthrough by observations in X-ray and other invisible radiation bands. Space exploration has revealed vital new information on the solar system. With the aid of space observatories the whole electromagnetic spectrum has been opened to view. In consequence we are witnessing the beginning of a new era in astronomical research.

The preparation of this edition has involved a major revision of the book which aims to provide an up-to-date account of the outstanding achievements of radio methods in the perspective of progress in other branches of astronomy.

It is now fifty years since the publication of the first discovery of radio waves from the Galaxy by K. G. Jansky of the Bell Telephone Laboratories, USA. This new edition is a tribute to the remarkable contribution radio astronomy has made to our knowledge of the universe during these fifty years.

J. S. HEY

Contents

Preface

IN THIS book I have attempted to give a readable survey of the whole field of radio astronomy. The greater part is devoted to the naturally emitted radio waves which enable us to observe astronomical phenomena at distances extending toward the limits of the observable universe; at closer range, within the solar system, radio echo methods can also be applied and are described in a chapter on radar astronomy. Radio observations have quickened the pace of progress in almost every branch of astronomical research, and have provided a great deal of fundamental information about astrophysical phenomena and the structure of the universe. Throughout the book I have first indicated the optical picture and then explained how radio has added to our knowledge.

The early history of radio astronomy is described in Chapter 1. The next chapter deals with the properties of radio waves and how they are generated by natural processes; readers who feel this is familiar ground or wish to proceed at once to the results of radio-astronomical observations may wish to omit Chapter 2. Similarly, Chapter 3 outlines the various types of radio telescopes.

The account of the results of research in radio astronomy starts with Chapters 4, 5 and 6 dealing with the solar system. The next two Chapters 7 and 8, are concerned with our Galaxy and the radio sources in the Galaxy. Finally, Chapters 9 and 10 describe other galaxies and the distant radio sources in the universe. The above groups of chapters essentially comprise complete sections of the book.

Although the book is designed for the intelligent layman, the account should also provide a useful survey for those scientists who look for an easily readable outline of progress in radio astronomy, and its contribution to our understanding of the universe.

J. S. HEY

1. Introduction

The Beginning of Radio Astronomy

IN 1894, six years after Hertz had first produced radio waves in the laboratory, Sir Oliver Lodge, believing that the radiation from the Sun extended far beyond the visible spectrum, was saying at a lecture before the Royal Institution of Great Britain, "I hope to try for long-wave radiation from the Sun, filtering out the ordinary well-known waves by a blackboard or other sufficiently opaque substance." Later he tried the experiment at Liverpool University where he was Professor of Physics. His attempts proved negative, and he concluded, "There were evidently too many terrestrial sources of disturbance in a city like Liverpool to make the experiment feasible. I don't know that it might not possibly be successful in some isolated country place, but clearly the arrangement must be highly sensitive in order to succeed." Sir Oliver Lodge clearly foresaw the possibility of radio waves from astronomical sources like the Sun, and also the necessity to choose interference-free observing sites.

Nordman in France was also unsuccessful, and in his report he referred to previous experiments by Wilsing and Scheiner in Germany. All these early attempts failed because electronic art and knowledge had not advanced sufficiently to ensure success. Ironically, when thirty years later radio waves from astronomical sources were detected, the discovery emerged incidentally from a study of atmospheric interference to radio communications.

The first observation of extra-terrestrial radio waves was made by Jansky at Bell Telephone Laboratories, USA. Jansky had constructed a rotatable aerial operating at a wavelength of 15 m to study the direction of arrival of atmospheric static that causes unwanted radio noise in communication receivers. By 1933 he had reached the conclusion that the received radio noise originated in three ways; local thunderstorms, the combined radiation of distant thunderstorms, and an extra-terrestrial source in the vicinity of the centre of the Galaxy. He subsequently demonstrated that the latter radio emission came from a source distributed throughout the Milky Way and with greatest intensity from the galactic centre. Jansky realised the implications of his discovery and suggested the construction of a parabolic mirror for further observations at metre wavelengths, but no support for his proposal was forthcoming.

The interest in radio astronomy was kept alive by the initiative of an American radio engineer, Reber, who in 1937 decided to pursue Jansky's

1

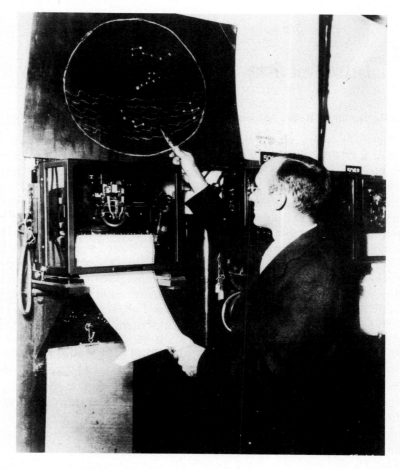

FIG. 1.1   The late K.G. Jansky of Bell Telephone Laboratories, USA, indicating the source of the cosmic radio noise he discovered.

discoveries further by constructing a parabolic reflector of 9.5 m diameter in his back garden at his own expense. This magnificent instrument, shown in Fig. 3.7, was the first radio telescope to be built for astronomical research. Reber started observations at a wavelength of 9 cm on the assumption that at short wavelengths he could achieve better resolution, and that if the radiation followed Planck's blackbody law the received power would be greater. Failing to detect the celestial radio emission at 9 cm and then at 33 cm, he was at last successful when he changed to a longer wavelength, 1.9 m. At this latter wavelength he obtained the first radio maps of the Milky Way. In addition to the main peak in the direction of

Sagittarius, he found minor peaks in Cygnus and Cassiopeia. Reber further suggested that radiation might originate from collisions between electrons and protons in ionised interstellar hydrogen.

The first recognition of radio emission from the Sun was made in 1942. In England, Hey (the author) was working as a civilian scientist in the British Army Operational Research Group directed by Sir Basil Schonland. As one of his research tasks Hey was responsible for analysing all occurrences of jamming of Army radar sets. For this purpose, a system for immediate reporting of the type and direction of jamming was organised, and radar operators were instructed in techniques of observing and recording interference. In February 1942, reports from many sites operating at 4 to 8 m wavelength led Hey to conclude that the Sun was radiating intense radio emission and that the phenomenon was associated with a large sunspot on the solar disk. Hey's report was restricted in circulation for reasons of military security. His conclusions seemed so unexpected that they did not at first receive unanimous acceptance by radio research scientists.

In the latter half of the same year, 1942, Southworth of Bell Telephone Laboratories made the first successful observations of thermal radio emission from the quiet Sun at centimetric wavelength. Again, the prevailing security regulations limited the circulation of the report. The two reports, by Hey on the intense radiation associated with solar activity, and by Southworth on the normal solar radio emission, laid the foundation to solar radio astronomy.

In 1944 a theoretical prediction proved to be another important landmark in radio astronomy. Reber's publications had attracted the attention of Oort at Leiden who suggested to van de Hulst that he should examine any possible astrophysical mechanisms that might produce radio spectral lines. Van de Hulst came to the conclusion that a line at 21 cm wavelength from neutral atomic hydrogen should be detectable. Since hydrogen is the major constituent of interstellar matter it was realised that observation of this line could be of great astronomical significance. Although it was not until 7 years later that the line was first observed, the prediction has proved well founded.

Further discoveries in new aspects of radio astronomy resulted from the research investigations initiated by Hey at the Army Operational Research Group. In 1944 when the threat of V 2 bombardment was imminent, Hey proposed a scheme for detecting the rockets as they approached England. The plan involved fitting new large aerial systems to Army radar sets and deploying them at ccastal sites. The scheme was readily accepted and went into operation within six weeks. In the first week of the V 2 action the only rocket detected was a wild shot that nearly hit the observing radar! After rectifying faults due to hasty installation, the approach of every subsequent V 2 falling in London or the surrounding area was observed. Two problems

arose during this operation. Transient echoes at a height of around 100 km gave rise to false warnings. Such echoes had long been known to ionospheric research workers as "short-scatter" echoes. Secondly, it was found virtually impossible to improve the radar receiver sensitivity because so much external radio noise was entering the aerial system. The unwanted noise was found to be predominantly the cosmic radio emission originally discovered by Jansky. The end of the war gave Hey and his colleagues Parsons, Phillips and Stewart the opportunity to make a scientific study of these phenomena. In 1945 they proved that the short-scatter echoes were radar reflections from meteor trails. They determined meteor radiants and velocities, and discovered daytime meteor showers. In 1946 they discovered the first discrete cosmic radio source, Cygnus A. In the same year a large sunspot enabled them to establish (in association with Appleton) the main radio properties of sunspots and solar flares. Radio emission from the Sun, the Galaxy, the discrete sources, and radar echoes from meteor trails all subsequently developed into important branches of radio astronomy.

The discoveries by Hey and his team drew the interest of J.A. Ratcliffe, to whom Hey, like so many others, was greatly indebted for his first introduction to radio science. Ratcliffe was returning to Cambridge after the war and he encouraged Ryle who had joined his staff in his plan to pursue radio astronomical research. At this time Lovell, who was resuming his research work on cosmic rays with Blackett at Manchester, believed that the radar methods used by Hey's team on meteor trails could be applied to cosmic ray showers. At Lovell's request, Hey and Parsons set up in Manchester a similar radar to their own. City interference from trams and factories impelled Lovell to move out to Jodrell Bank, a site owned by the botanical department of the University. There he realised that the cosmic ray showers were too elusive to capture by radar and turned his attention instead to the study of meteor echoes, and later to other branches of radio astronomy. From these beginnings two famous university radio astronomy research centres became established, under Sir Martin Ryle at Cambridge and Sir Bernard Lovell at Manchester. At the same time the initial discoveries had also stimulated the interest of the late J.L. Pawsey of the Commonwealth Scientific and Industrial Research Organisation, Australia, and with Bowen's encouragement Pawsey set up an equally well-known radio astronomy group there.

It is impossible to summarise even briefly the highlights of subsequent achievements in radio astronomy. The progress that has been made is described in the following chapters of this book. A few of the early results will be noted here. In 1946, natural thermal radio emission from the Moon was detected by Dicke and Beringer in the USA, and radar echoes from the Moon were obtained by the US Army Signal Corps and by Bay in Hungary. In 1948, Bolton in Australia made the first positive identification of a discrete radio source with a visual galactic object, the Crab Nebula;

and F.G. Smith developed very accurate position finding methods at Cambridge leading to the identification of Cygnus A in 1951 with a very distant galaxy. The 21 cm hydrogen line was also first observed in 1951, and the astronomical potentialities of this spectral line were most successfully exploited in studies of the Galaxy by the Leiden and Australian groups.

After van de Hulst's prediction of the 21 cm hydrogen line, the most important early theoretical advance was made by the Swedish astonomers Alfvén and Herlofson, and the Russian astronomer Shklovsky in establishing the synchrotron process as the origin of most of the intense continuum radio emission prevailing in different types of astronomical sources.

Radio astronomical research observations are now established in many places throughout the world. The first discoveries by Jansky were made in USA, although for a long time interest was only kept alive in that country by Reber. Later, after the subsequent discoveries in Britain, real support for radio astronomy in USA was forthcoming. Important research centres were also set up in many countries, for example, Australia, Holland, France, Canada, Russia, Japan and so on.

One of the amazing attributes of radio astronomy has been its continuing contributions to the fund of unexpected phenomena. To mention only two in recent years, the discovery of quasars and pulsars. Perhaps the most important development of all, however, is that the radio methods have become an integral part of astronomy, combining with optical and other methods of observations to lead to a deeper understanding of the nature of the universe.

In the next chapter I shall discuss the nature of radio waves and how they are generated. The following chapter describes the various kinds of radio telescopes. These Chapters 2 and 3, together with the Appendices, cover in a fairly elementary way the fundamentals of radio applied to astronomy, so making the account complete. Readers who prefer to continue at once with the description of the results achieved in radio astronomy should proceed to Chapter 4.

2. Radio Waves

Comparison between Light and Radio Waves

In this chapter we are going to consider the following questions. What is a radio wave? How can radio waves be produced naturally in the universe? Before we discuss radio waves let us first consider the properties of light waves. Familiar objects have such obvious reality to us that we often forget that we only know them by certain responses on special types of detectors that we happen to possess. We recognise a candle, for example, by the light it emits, and a table by the light it reflects or scatters. Scientific investigation has shown that the light actually consists of electrical waves of very short wavelength. The wavelength is simply the distance from the crest of one wave to the next. We have the natural ability to detect, that is, to see the radiation of very short wavelength that we know as light. These visible radiations have wavelengths extending from 0.4 to 0.8 μm. Different parts of this wavelength band appear to us different colours. The longest wavelengths appear as red, and with decreasing wavelength we pass through the other colours of the spectrum, orange, yellow, green, blue, indigo and then violet, the shortest visible wavelength. If the whole range of wavelengths are present together we see the radiation as white light. An object like a pillar box looks red because it strongly reflects the red wavelength, while absorbing the others. The human eye has the ability to distinguish the different bands of wavelengths, or to say the same thing in scientific terms, to analyse the spectrum of the observed radiation.

The eye can also separate the light coming from different directions. Even objects as close as 1 minute of arc (1/60 of a degree) can be distinguished. The angle in the diagram is 10°, so an angle of 1 minute of arc (1') is 1/600 part of this. We use this ability to see the detail of a scene, or to read small print. In scientific language, we say the eye has good resolving power.

Now ordinary objects such as candles or tables (and even people) both emit and reflect radio waves. These are also electrical waves, similar to

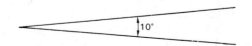

FIG. 2.1. A minute of arc (1') is 1/600 part of this angle.

light waves, the only real difference being that the wavelength of radio waves range from a millimetre to many metres. As compared with light, radio waves at 50 cm wavelength, for example, are about a million times longer in wavelength. As we have no natural ability to detect such wavelengths directly, we are not normally aware of the presence of this kind of radiation. We therefore require an intermediate detecting instrument. We must build a piece of apparatus, a radio receiver, that converts radio waves into an output we can see and hear. If we wish to make measurements rather than just listen, we must display the output as a dial reading, or on a chart.

Now you may ask, if ordinary objects emit radio waves why do we not hear them on our radio sets at home? The answer is that it requires extremely sensitive radio equipment to detect the radio waves emitted naturally and continually by surrounding objects. And if you listened to the detected waves on a loudspeaker, by amplifying them sufficiently to make them audible, they would only sound like the background hissing noise you can sometimes hear faintly on a radio set.

However, certain natural phenomena are easy to detect because they emit radio waves strongly. For instance you often pick up radio waves from lightning flashes, and a short-wave radio or a television set is quite capable of detecting the radio emission from active sunspots or the radio outbursts from Jupiter. Unfortunately, there are usually so many artificial sources of radio waves in towns from electrical devices and machinery causing radio interference that it is not easy to sort out natural sources without making specially designed experiments. We shall discuss the types of apparatus used by radio astronomers in Chapter 3. Just to give an idea of how weak are the radio waves detected from astronomical sources, it has been calculated that all the energy picked up by radio telescopes in the last twenty years would only be sufficient to heat a teaspoonful of water about a millionth of a degree!

The other requirement in radio astronomy is to be able to look in different directions, and distinguish one source from another. As we shall show later, in order to have good resolving power the radio telescope must be very many wavelengths in size. The wavelength of light is so small that the human eye is millions of light wavelengths across. But radio wavelengths are so long that even a radio telescope as big as that at Jodrell Bank cannot compete with the resolving power of the eye. At metre wavelength we should need a radio telescope about two miles in diameter to have the same ability as the eye for seeing detail. How radio astronomers have overcome this problem will be explained in Chapter 3.

If we study the various properties of radio waves we can demonstrate that they are really similar to light waves, except for the difference in wavelengths. For instance, radio waves and light waves travel through space at exactly the same speed of 186,000 miles a second. In the metric

system, this is equal to 300,000 km/sec. A simple way to demonstrate that both radio and light radiation consists of waves is to let part of the radiation follow one path and the other part another route, and then bring the two together again. If the radiation is a wave motion, the two parts may come together in step and be observed with maximum intensity, or they may be out of step and cancel each other out. This is just what happens with both light and radio. One way of carrying out this experiment is by combining a direct ray with another reflected at a glancing angle from a mirror, as shown in Fig. 2.2. The method was first applied to light by Lloyd in 1837.

FIG. 2.2.   Lloyd's mirror experiment.

The mixing of the two waves is often called wave interference. The interaction between radio waves travelling along two different paths to a radio receiver is the usual cause of periodic fading of radio signals we experience in listening to a distant radio broadcast.

We can represent any kind of wave motion by Fig. 2.3 shown below. In a water wave the diagram represents the up and down displacement of the water. In an electrical wave, however, the diagram is used to illustrate the varying amplitude of the electric field. Any type of wave has a speed, frequency and wavelength which are related to each other very simply.

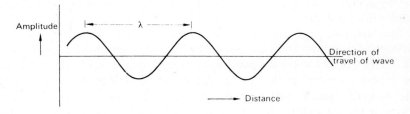

FIG. 2.3.   Wave diagram.

In the diagram, the distance from one wavecrest to the next is the wavelength λ. The number of waves passing any point each second is called the frequency, f. So the distance travelled by the wave in 1 second is $f\lambda$. But the distance moved per second is the wave velocity, c. Hence we have $c = f\lambda$. The equation enables us to calculate the frequency if we know the

wavelength, and vice versa, because we know that the speed $c = 3 \times 10^8$ m/sec. Radio wavelengths are generally expressed in metres, and frequencies in megacycles per second, that is 10^6 c/s, usually written Mc/s.

Thus we have $f = \dfrac{c}{\lambda} = \dfrac{3 \times 10^8}{\lambda}$ c/s $= \dfrac{300}{\lambda}$ Mc/s.

Hence if $\lambda = 10$ m, then $f = 30$ Mc/s, or if $\lambda = 10$ cm $= 0.1$ m, then $f = 3000$ Mc/s.

It is becoming quite usual nowadays to write hertz instead of cycle per second. The abbreviation is Hz in place of c/s. This terminology is taken from the name of Hertz, the German physicist, who in 1888 first produced radio waves in the laboratory.

The apparatus used by Hertz is illustrated in Fig. 2.4.

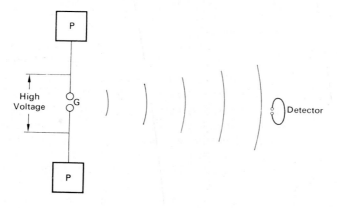

FIG. 2.4. Demonstration of radio waves by Hertz.

A high voltage produced by an induction coil caused a spark at the gap G. The surge of current due to the spark oscillated between the two plates P. Radio waves about 5 m wavelength were generated in this way. Hertz detected the radio waves by means of a metal circle with a very narrow gap. Provided the detector was not too far away, the received waves produced enough voltage to make tiny sparks appear across the detector gap.

We have described radio (and light) as electrical waves, but they are more correctly called electromagnetic waves. This is because electrical changes are always accompanied by magnetic fields. One can demonstrate the magnetic effect of a direct electric current very easily by having a magnetic compass needle near a wire and then connecting the wire to battery terminals. The compass needle is deflected in a direction at right angles to the current flow.

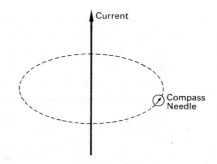

FIG. 2.5.   Magnetic effect of a current.

If you carry out this experiment near a sensitive radio receiver you may even hear the click because a radio wave is emitted when the current changes as the wire is connected.

We are most familiar with the waves we see on the surface of water. Another type of wave is produced by vibrating the end of a long rope or wire. All wave motions involve exchanges between two kinds of energy. In a water wave, for example, the water moves up and its weight makes it fall back again. The exchange is between energy of motion and the gravitational energy associated with weight. The disturbance at one place affects the adjoining water and so gives the appearance of a travelling wave. In a vibrating wire or rope, energy is transferred between motion and tension. In an electromagnetic wave the exchange is between electrical energy and magnetic energy.

It is interesting that the speed of an electromagnetic wave was first calculated theoretically by Maxwell in 1873. Because the wave motion depends on the exchanges between electrical and magnetic fields, he deduced that the speed is simply the ratio of force expressed in magnetic units to that expressed in electrical units, and the value comes precisely the same as the measured velocity of light and radio waves.

Electrons

Before we proceed further it is worth emphasising that science is no more than a description of behaviour. The aim of science is to describe as many phenomena as possible in terms of basic concepts and laws. Remembering the descriptive role of science makes it much easier to accept some of the less tangible concepts when we realise that they have been built up as ideas to describe the way things happen. From observations, we deduce by logical argument what we think must be the basic constituents and rules of action between them. It is by this process that atomic theory and our

knowedge of elementary particles has been derived, and our ideas about electric and magnetic fields have been developed.

The scientific concepts have their starting point in very simple experiments. For instance, the initial ideas of electrical charges follow by first showing that a piece of ebonite or plastic rubbed on a cloth picks up small bits of paper. We say the substances are electrified, and we proceed with other experiments to find more precisely how electric charges behave. We find that there are two kinds of electrical charge which we decide to call positive and negative because when they are combined they neutralise each other. We also find that a negative charge is attracted to a positive charge, and we describe the force as due to an electric field.

The laws of chemical reactions enable one to conclude that substances are composed of elements and the smallest particles are atoms, which may combine together to form molecules. Further experiments prove that the atoms consist of lightweight negatively charged particles called electrons and positively charged ions. The lightweight electrons can easily be knocked out of an atom by various stimuli, for instance by a collision with a fast moving particle, or by ultraviolet radiation. The atom is then said to be ionised. In the case of hydrogen, the positive ion is called a proton. Hydrogen is particularly important in astronomy because it is the predominant constituent element of the universe. The mass of the electron is $\frac{1}{1840}$ that of the hydrogen atom. As the electrons are so light their motion is easily altered in an electric field. As we shall see, radio wave phenomena are intimately connected with the motions of electrons.

When an electron changes its motion, the disturbance generates a radio wave, just as a stone thrown in water causes a disturbance that starts a wave on the surface. If an electron is made to accelerate or decelerate, then the electric field produced by the electron undergoes a corresponding variation, and an electromagnetic wave is consequently propagated out from the electron. If we can keep an electron oscillating, so that it is periodically changing its motion at a certain frequency, then radio waves with an oscillating field will be emitted at that frequency. Such waves are produced at a radio transmitter by applying an oscillating voltage to the transmitting aerial, the electrons in the aerial being made to surge backwards and forwards at the transmitter frequency.

Thermal Radio Emission

We said earlier that solids emit radio waves, and with sensitive apparatus we can detect the radio waves emitted, say, from a table or a rock. Not only are these substances very poor conductors of electricity, possessing few free electrons, but we are also stating that solids emit radio waves

when no voltages are applied. The explanation is found in the continual state of vibration which is really the heat energy of the substance. Heat is a form of energy and consists in the motion of constituent particles, and the higher the temperature the more rapid is the motion. Although the molecules in a solid are bound together, their heat energy causes them to vibrate. In consequence, the positive and negative charges (ions and electrons) are always being subjected to slight displacement. The movement of electric charges produces the natural thermal radiation from a solid. This radiation may be very weak but occurs over a very wide range of wavelengths, including radio wavelengths. Of course if the object is very hot, the visible radiation becomes obvious as the object begins to glow, first red, and then white as the temperature is raised. The intensity of radio emission that may be expected at different wavelengths, that is, the radiation spectrum of a perfect radiator, was derived theoretically by the German physicist Planck in 1901, and his conclusions were in excellent agreement with the experimental measurements previously made by Lummer and Pringsheim in 1897. The dependence of radiation intensity on wavelength is shown in Fig. 2.6.

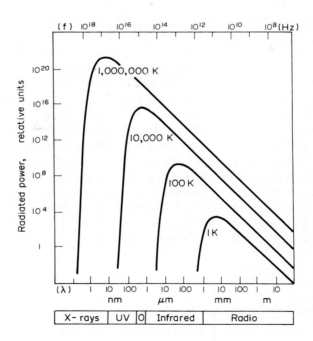

FIG. 2.6.   Spectrum of a perfect radiator. The optical band is marked 'O'.

The maximum intensity for any given temperature appears at a wavelength λ_m given by a simple formula known as Wien's Law, namely

$$\lambda_m T = C$$

where C is a constant value, approximately equal to 0.5 where λ is in centimetres and T in degrees (K) above absolute zero.

For a white hot body at 6000 K, like the surface of the Sun, the maximum intensity is at the wavelength of yellow light. For an object at normal room temperature, 17°C, equal to 300 K, the total radiation is of course very much less while the maximum intensity occurs at an infrared wavelength about 15 μm (micrometres). As Fig. 2.6 shows, at temperatures of 1 K or less the emission is weak and mainly confined to radio wavelengths. At temperatures of a million K or more the intensity is far greater, with X-rays radiating most power. At 10,000 K the maximum is in visible radiation. For all temperatures above 100 K the tail of the spectrum extends to include radio waves. Thermal radio waves usually lie in this long wavelength spectral region.

For these long wavelengths, the intensity of radiation was derived in 1900 by the British physicists Lord Rayleigh and Sir James Jeans. The Rayleigh–Jeans law states that the radiation power flux P from a perfect radiator at absolute temperature T is given by

$$P = kT/\lambda^2.$$

The constant k is known as Boltzmann's constant.

More precisely, P is the power flux on unit area per unit frequency bandwidth per unit solid angle for one polarisation. We may note here that we shall generally refer to the free-space wavelength λ rather than the frequency since the wavelength is so easily expressed in the range of metres to millimetres.

Let us pause for a moment to consider what is meant by a "perfect" radiator. There is a well-known principle that if an object is in equilibrium with its surroundings, for example, an object at normal room temperature (say 17°C) in an enclosure at the same temperature, it takes in just as much heat as it loses. Obviously so, for the object will not get hotter or colder if the surroundings are all at the same temperature. It follows that an object which absorbs a lot of radiation must also emit a lot. If it absorbs all the radiation and is therefore a perfect absorber, it must emit equally well and consequently be a perfect radiator. On the other hand, if radiation falls on a material partially transparent to the radiation much of it goes straight through. The material radiates only as much as it absorbs.

Thermal Radiation from a Gas

In radio astronomy we may observe the radiation from solid surfaces such as the Moon and planets, but more often we are concerned with the radiation from the gaseous atmospheres of stars and galaxies. The hotter the gas, the faster the particles move, and the more furiously do they collide with each other. The first effect of this is to break up the molecules into atoms. If the gas is very hot, then electrons are knocked off the atoms, which therefore become electrons and ions. These negative electrons and positive ions share in the random motion and repeatedly come close to each other as they move about. Now when an electron moves near a positive ion a force will be exerted on it since opposite charges attract each other. The electron will be deflected in its path by this force as shown in Fig. 2.7.

Moving Electron

⊕ Ion

FIG. 2.7.   Path of electron deflected by a positive ion.

The change of motion generates a radio wave. Some of the electric field of the electron is thrown off as it accelerates round the bend and is radiated out into space. The radiation emitted in this way is the thermal radio emission from the gas, because it arises as a result of the thermal motion.

There are several possible causes of ionisation. For instance, if a gas becomes hotter the atoms move at higher speeds and the fastest may knock out electrons simply by collisions between them. Very often in astronomy, the ionised gas is found in extensive regions surrounding hot stars, and the ionisation mainly arises from ultraviolet and X-ray emission causing the ejection of electrons from atoms.

Let us now consider what happens to radio waves passing into an ionised gas. The electric field of the incident radio wave accelerates the electrons and makes them oscillate to and fro at the radio frequency. As the electrons vibrate they, too, will send out a wave at the radio frequency. This secondary wave combines with the incident wave and slightly alters its path. The bending of the direction of the wave is called refraction. If there are enough electrons in the ionised gas, the secondary wave is so strong that total reflection is produced. The longer wavelengths are affected most. A familiar example is the distant reception of long wave broadcast radio by reflection from the ionosphere. For any given electron density, that is, the number of electrons per unit volume, total reflection occurs when the radio frequency excites the natural frequency of oscillation of the electrons. This

frequency is called the critical frequency, or the plasma frequency. Plasma is simply another name for an ionised gas. The more electrons the higher is the critical plasma frequency. A simple approximate formula is that the plasma frequency is $f = 9 \sqrt{N}$, where N is the number of electrons per cubic metre.

We must now consider how radio waves can be absorbed by an ionised gas. We will suppose the frequency to be high enough for the radio waves to be able to travel through the ionised gas. The electrons of the gas oscillate due to the action of the radio wave. If the number density is sufficiently great, the electrons cannot oscillate freely without colliding with gas atoms and ions. The jostling due to frequent collisions will make the electrons' motion tend to become random, so the energy they have acquired from the radio wave appears as thermal motion and thermal radiation. The energy of the incident radio wave has therefore been absorbed and turned into heat energy.

As we explained earlier, a substance that is a perfect absorber is also a perfect radiator, and the intensity of radiation is given by the Rayleigh-Jeans Law, $P = kT/\lambda^2$. The easiest way of assessing how much thermal radio emission will be radiated by an ionised gas, is by considering what fraction of incident radiation it absorbs. If an ionised gas absorbs a tenth of the radiation falling on it, then we know it must emit a tenth that of a perfect radiator, so that $P = 0.1\ kT/\lambda^2$.

We can understand how radiation is related to absorption or opacity quite easily by thinking of the partly transparent gas flame of a gas fire. To make the fire radiate efficiently the fireclays are put in the flame. This is a solid opaque substance and a near perfect radiator at the temperature of the flame. In comparison, the flame itself is a poor radiator because of its transparency.

If we have an ionised gas that is semi-transparent to radio waves we can say then that the radiation intensity is

$$P = \epsilon \frac{kT}{\lambda^2}$$

where we regard ϵ as an emission factor depending on the absorption of the gas. If the gas becomes opaque and absorbs the radio waves, then $\epsilon = 1$.

In a semi-transparent gas, ϵ is proportional to λ^2, thus cancelling out the term $1/\lambda^2$ in the expression for P. The intensity of radiation in this case is independent of the wavelength. The electron density and the temperature influence ϵ because they affect the rate of collisions. It follows that for a semi-transparent gas, the intensity of radiation depends only on the number of electrons and temperature.

We can illustrate the spectrum of thermal radio emission from an ionised gas at a given temperature by Fig. 2.8.

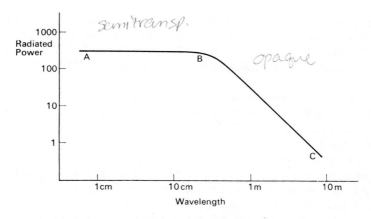

FIG. 2.8.   Spectrum of thermal radio emission from an ionised gas.

The part AB of Fig. 2.8 corresponds to short wavelengths when the gas is semi-transparent. At long wavelengths the gas becomes opaque to radio waves, and the spectrum given by $P = kT/\lambda^2$ is the part BC of the graph. We know that observed radio emissions are thermal if they fit this type of spectrum and if the intensity corresponds to possible values of the temperature.

Non-thermal Radio Emission

The radio emission from some solid bodies, like the Moon and most planets, is thermal. Thermal radio waves can be detected from ionised hydrogen surrounding the Sun and hot stars. But in many astronomical objects we find that thermal radiation is swamped by far more intense sources having spectra that cannot possibly be thermal. In fact, a common type of spectrum is shown in Fig. 2.9.

Comparing this with Fig. 2.8, we see that the spectrum is not that of thermal radiation. This spectrum can be described mathematically by saying that the power radiated is proportional to λ^x where x is called the spectral index and has an average value of about 0.6.

Two main criteria for judging when radio emission is not thermal are therefore the exceptional intensity and spectrum. The term "radio brightness" is often used to describe the radio power flux received per unit solid angle from a source. Even for a non-thermal source, it is often convenient to refer to the "radio brightness temperature" derived by the formula $P = kT_B/\lambda^2$. An abnormal value of T_B shows at once that radiation is not thermal. We must remember in this case that T_B is not a true temperature, but merely a convenient way of expressing the intensity.

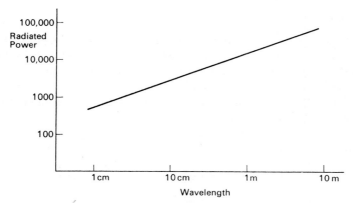

FIG. 2.9. Typical spectrum of non-thermal radiation.

There are several processes, both natural and man-made, which may produce non-thermal radio emission. The transmitters used for broadcasting or in radar equipment radiate such very high powers from relatively small aerials (many kilowatts or even megawatts in short pulses) that if we tried to work out their brightness temperatures we would find them to be quite fantastic (many million million degrees). It is obvious that the radio power in these cases is not generated thermally! It is in fact usually produced by a circuit making an electric current surge backwards and forwards at a chosen frequency. This oscillating current may be fed to a dipole consisting of a metal conductor half a wavelength long. The electrons oscillate up and down the dipole and so produce a radio wave moving out into space.[1] The electric field of the radiated wave is illustrated in Fig. 2.10 which shows two different ways of representing the wave.

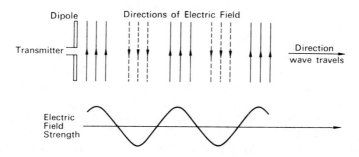

FIG. 2.10. Wave radiated from a transmitting dipole.

[1] When electrons oscillate in unison the radio waves add together and are said to be "coherent". Random motions, as in thermal radiation processes, are described as "incoherent".

Wave Polarisation

We notice in Fig. 2.10 that the electric field of the wave is parallel to the length of the dipole, and for this reason the wave is said to be linearly polarised. On the other hand, the electrons producing thermal radiation in a gas or solid move at random in all directions. Consequently, the electric field in thermal radio emission occurs in random directions, and the wave is said to be randomly polarised.

Another possible kind of polarisation is circular, a type of wave radiated by electrons moving in circles, which occurs in the presence of a magnetic field.

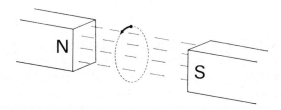

FIG. 2.11.   Electron circles round lines of magnetic field.

Such motion is illustrated in Fig. 2.11 showing the electron moving at right angles to the field and so going round in circles like an electric motor. If the electron also has a component of velocity along the field then it twists in a circle as it travels forward, so that it gyrates in a spiral as shown in Fig. 2.12.

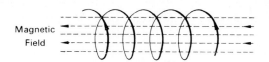

FIG. 2.12.   Spiral motion of electron in magnetic field.

It is the guiding of electrified solar particles in spirals along the magnetic lines of the Earth's field towards the North Magnetic Pole that causes the displays of Northern lights (Aurora Borealis) and to the South Pole (Aurora Australis).

An interesting consequence of the circular motion is that the electrons radiate a wave with a rotating electric field, that is a circularly polarised wave, as illustrated in Fig. 2.13(a).

Of course if we view the circle from the side it appears as a line as in Fig. 2.13(b), and the radiated wave in this direction is linearly polarised.

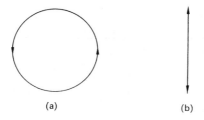

(a)   (b)

Fig. 2.13.   Polarisation from electron in circular motion. (a) An electron circling about a field perpendicular to the paper. The electron radiates a circularly polarised wave towards us. (b) The circle viewed end-on looks like a line. The wave from the electron viewed in this direction is linearly polarised.

We can say that if we find naturally emitted radio waves that are circularly polarised then a magnetic field must be present. Also, if we look at right angles to the magnetic field, that is end-on to the electron circles, the field is likely to be linearly polarised.

Just as a pendulum swings with a natural frequency of oscillation, so an electron has a characteristic frequency of gyration in a magnetic field. This is known as gyro-frequency, and is equal to 28 H in MHz where H is the magnetic field measured in mT.

We have seen that a magnetic field can cause the polarisation of radio waves. As we shall describe later, radio emission from astronomical sources often shows evidence of polarisation. This is not surprising when we realise how prevalent magnetic fields are throughout the universe. We know for example, that the Earth has a magnetic field, while sunspots have very strong fields. Magnetic fields occur in many astronomical objects, and weak magnetic fields even pervade interstellar gas.

Faraday Rotation

Suppose linearly polarised waves are generated by a distant radio galaxy. Along the path from the source to the observer the radiation is likely to encounter ionised gas and magnetic fields, particularly in the vicinity of the source and within our Galaxy. The electrons and magnetic field along the path have the effect of rotating the direction of the electric field of the wave, or as we say, rotating the polarisation. This is known as Faraday rotation, because this kind of action was first discovered by Michael Faraday when he passed polarised light through glass in the presence of a magnetic field.

We can understand the rotation of polarisation in the following way. The linearly polarised wave makes the electron vibrate parallel to the electric field of the wave. At the same time the steady magnetic field that is present is trying to make the electron go round in circles. As a result the

polarisation of the wave is twisted. The amount of rotation clearly depends on the total number of electrons in the path and the strength of the magnetic field. The amount of Faraday rotation is less the shorter the wavelength. By observing the direction of polarisation at different wavelengths, it is possible to derive valuable information about electron densities and magnetic fields encountered along the path of the wave. Such studies also enable us to unravel the initial direction of polarisation of the emitted wave, and hence deduce the orientation of the magnetic field at the source.

Synchrotron Radiation

We have yet to explain why many astronomical sources emit enormous power at radio frequencies, and where the energy comes from. In 1951, two Swedish scientists, Alfvén and Herlofson, suggested that the really intense radio emission might arise from electrons with enormously high velocities, like those found in cosmic rays, radiating as they spiral in magnetic fields. No one is very certain how cosmic ray particles acquire such great speeds, but we do know that they exist. The speeds are so high that they are not far from the velocity of light. If an electron is made to change its direction by a magnetic field it will radiate, but as the electrons are going so fast the radio waves are pushed out in the forward direction. This means that an observer only receives radiation from an electron moving towards him as shown in Fig. 2.14.

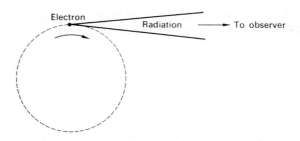

FIG. 2.14.   Beaming of radiation from high-speed electron.

As the circle is end-on to the observer the received radiation will be linearly polarised. When it was worked out how much energy could be radiated it was found to be very large for electrons with very high velocities, like those of cosmic rays. This process can explain the intensity and spectrum of the non-thermal radio emission so often observed in radio

astronomy. The reason why the process is so effective is that the particles have such a tremendous store of energy because of their high velocity.

Radiation produced in this way was first recognised in a laboratory machine called a synchrotron, designed to give electrons tremendously high speeds for use in laboratory investigations of collisions between fast particles and atoms. In this machine, electrons circled in a magnetic field and they were seen to emit a small amount of light. The name synchrotron emission is now given to radiation produced in this way by high energy particles in magnetic fields. It is more easily generated at radio wavelengths because less energy is required.

Synchrotron radiation is emitted over a very wide frequency band, which can be explained in the following way. If an electron gyrates slowly in a magnetic field, any radiation is closely confined to the gyro frequency (28 H in MHz) or to simple harmonics of this frequency. When the electron moves at speeds near the velocity of light the circumstances are altered. Firstly, the gyro-frequency changes because the mass of electron is increased by its high velocity in a manner predicted by Einstein's theory of relativity. For this reason, particles with speeds comparable to that of light are called "relativistic". The other effect of high speeds is to spread the spectrum, which results from the "beaming" of the emission in the forward direction. The beaming means that the observer only sees sharp pulses of emission at those instants when a gyrating electron happens to be directed towards the observer. Since the spectrum of a short pulse always contains a wide range of frequencies, synchrotron radiation covers a wide frequency band. In astronomical sources there will be vast numbers of electrons with a great range of energies, so that the radio emission we observe will be due to the combined effect of all of them.

Calculations show that much of the non-thermal radio emission from astronomical sources can be explained if the magnetic field is only 1 to 10 nT, that is less than a ten thousandth of the strength of the magnetic field at the Earth, even if there is only a quite small density of electrons, provided they have high energy. It is usual to describe the energies in terms of electron volts (eV). The electron energies required to produce the observed radio emission would be about 10^9 eV. This means that the electrons have the energies that they would gain if they were accelerated in a potential of 10^9 volts, that is, a thousand million volts, making them acquire speeds close to that of light.

Not only can we account for the observed high intensities of radio emission in this manner, we can also quite easily explain the shape of the spectrum. All that is necessary is to make the reasonable assumption that there are fewer electrons of higher energies, giving an energy distribution of the electron similar to that found in cosmic ray particles. We can then derive a spectrum of radio emission by the synchrotron process with an intensity proportional to λ^x, where x is somewhere between 0.2 and 1.2, so

fitting the observed spectrum of non-thermal radio emission. Radio emission by this process is now generally accepted as the correct explanation of the intense radio emission from many radio sources. As we shall see later, it was the Crab Nebula radio source within our own Galaxy that gave the first decisive evidence of the synchrotron type of radiation from an astronomical object. In the Crab Nebula, the linear polarisation of the optical and radio emission indicate that both are generated by the synchrotron mechanism.

It is especially notable that very low concentrations of electrons are adequate to produce the observed synchrotron radiation. For instance, an average density of 1 electron per million cubic metres is sufficient to account for the radio emission from the Galaxy. The reservoir of energy resides in the kinetic energy of the electrons at such extreme velocities and in the magnetic field together with the vast extent of the source.

We must now consider the length of time that an electron can radiate before its energy becomes exhausted, and how this lifetime depends on the type of radiation emitted. Compared with radio, higher electron speeds in stronger magnetic fields give rise to optical and X-ray synchrotron emission. Highest power is generated at a frequency proportional to HE^2 where H is the magnetic field strength and E the electron energy. A field $H = 1$ nT with $E = 10^9$ eV yields a maximum emission of about 160 MHz ($\lambda \sim 2$ metres). For the same field strength, electron energies would need to be of the order of 10^{12} eV to generate optical radiation, and around 10^{14} eV for X-rays. To illustrate how close to the velocity of light are these electrons we may note that an energy of 10^9 eV implies a speed of 99.99999 per cent that of light! Although the precise process of acceleration to such enormous speeds remains uncertain the prevalence of highly energetic particles in astrophysical phenomena is evident. For example, even the Sun ejects cosmic rays during solar outbursts.

The lifetime of relativistic electrons is controlled by the rate they lose energy by radiation. Rapid loss leading to a short lifetime requires replenishment of energetic electrons if the radiation is to continue. It can be shown that a relativistic electron radiating synchrotron emission loses half its energy in $835/H^2E$ years where E is the electron energy in MeV and H the magnetic field in mT. For radio, the lifetime may be so long in some cases that the energetic electrons may have originated in a single explosive event. For instance, an electron energy of 10^9 eV (1000 MeV) in a field of 1 nT means a lifetime of nearly 10^8 years. But in stronger fields the lifetime is shorter, and especially so at the higher electron energies generating optical and X-ray synchrotron emission. For example, if optical synchrotron radiation is produced by electrons of 10^5 MeV in a field of 0.1 mT the electron lifetime is no more than 80 years. Much shorter lifetimes are involved when X-rays are emitted.

When synchrotron radio emission is very intense, reabsorption may

ensue and dramatically reduce the radiation at longer wavelengths. The radio brightness at any wavelength cannot exceed "perfect radiator" emission at the equivalent temperature corresponding to the electron energies. Self-absorption particularly affects very powerful compact sources. It is recognised by a characteristic sharp fall in the spectrum at long wavelengths. It has proved to be a valuable indicator of the sizes of small but very powerful sources.

The Compton Effect

Another effect which can influence high intensities depends on interaction between electrons and radiation. To understand the interaction we have to realise that electromagnetic waves also behave like corpuscles which can collide with electrons. Classical physical ideas were revolutionised in the early part of the century by concepts introduced by Planck and Einstein demonstrating that radiation can act like particles, called photons, travelling with the speed of light. In consequence, when a photon collides with an electron an exchange of energy occurs. The electron acquires a velocity of recoil, while the photon's loss of energy is revealed by a change to lower frequency. The effect was first demonstrated by Compton in 1922.

In the synchrotron process, the electrons producing radio emission have speeds approaching that of light. In such circumstances the electrons can impart energy to photons during collisions. This is the "inverse Compton effect"—the colliding electrons lose energy and the photons acquire a higher frequency.

The transference of energy from relativistic electrons due to the inverse Compton effect leads to two important consequences. One is that the loss experienced by the electrons of intense sources becomes so considerable that it sets a limit of about 10^{12} K to the maximum radio brightness temperature that can be achieved. The limit provides a useful guide in the theoretical interpretation of intense radio emission. If the radio brightness much exceeds 10^{12} K it follows that the radiation emanates from a coherent process and not by the normal synchrotron mechanism. Another consequence of inverse Compton scattering occurs where there is sufficient density of high energy electrons and radiation photons. The relativistic electrons colliding with the photons may impart enough energy to the photons to convert them to optical radiation or X-rays.

Spectral Line Radiation

A different type of radio emission is spectral line radiation, which means that it is concentrated close to one wavelength. This type of radiation is well known at optical wavelengths. Line radiation occurs, for example, when common salt is vapourised by a fire or gas flame. An intense yellow

light is emitted by the sodium atoms in the salt, and if the light is examined by a spectroscope it is found to be concentrated at a particular wavelength, and the radiation is known as the sodium line. The gas discharge lamps for street lighting emit lines characteristic of the gases in the lamps.

The quantum theory applied to atoms accounts for spectral line radiation. In 1901, Planck deduced that all energy changes occur in small steps. The Danish physicist, Niels Bohr, subsequently showed that atomic spectra can be explained if atoms exist only in certain particular energy states. When an atom absorbs energy it changes to a high energy state. It can release the energy again by emitting radiation and jumping back into the lower state. In making this transition the atom radiates a quantity of energy (known as a quantum) which is related to the frequency of the radiation by Planck's Law:

$$\text{Energy change} = \text{Planck's constant} \times \text{frequency} = hf$$

where h denotes Planck's constant.

The transition between particular energy states explains why the radiation appears only at certain frequencies. As the frequency of radio waves is about a millionth that of light waves, the energy change involved in producing a radio wave is correspondingly small. So if we are to find radio lines from astronomical sources, there have to be large numbers of atoms or molecules in appropriate states for transitions to occur with energy changes corresponding to radio wavelengths.

The element that is found in greatest abundance throughout the universe is hydrogen, which possesses the simplest of all atomic structures consisting of a proton with an electron moving round it. The laws of quantum mechanics tell us that only certain orbits are allowed for the electron. The closest electron orbit has the lowest energy and is known as the "ground state". If the hydrogen atom absorbs energy by a fast collision, or by being exposed to radiation of sufficiently high frequency, the electron is pushed out into a higher energy orbit. The atom is then described as being in an "excited" state. When the electron jumps back it emits radiation. All transitions in hydrogen down to the ground level have too much energy for radio waves to be generated. In fact, such transitions for hydrogen all produce ultraviolet radiation. The excited electron in a high level state may instead jump back to one state above ground level. If it does this, the transition produces a light wave. If the excited electron returns to one state higher still, the infrared waves are generated. The less the energy change involved, the longer the wavelength. But we have to look for transitions in outer levels, for example, from the 91st to 90th level before we find any close enough to produce radio waves in this way. Although the chance of such outer-level transitions is small, Russian astronomers first demonstrated theoreticaly and then experimentally that these transitions can be observed in ionised regions near hot stars. Here the rate of ionisation is

balanced by recombination, and the radio spectral lines are emitted as the recombining electrons cascade down through the outer levels.

Much of the interstellar gas consists of neutral hydrogen atoms in the ground state. Now as it happens, there is one particular transition within the ground state, called a hyperfine transition, that causes the internal energy of the hydrogen atom to change by a small amount, corresponding to a radio quantum of 21 cm wavelength. This hyperfine transition is due to the fact that both the positive nucleus, the proton, and the electron in the surrounding orbit are spinning about their axes. These spinning particles act like two tiny electromagnets because as we have seen, moving charge constitutes a current, hence producing a magnetic field. If the two spins are in the same direction the atom has slightly more energy than when they are in opposite directions. When an electron in the higher energy state changes its direction of spin the hyperfine transition occurs and the radio quantum is emitted. The process is illustrated in Fig. 2.15. The probability of a natural transition from the higher energy state is so low that the average rate for an atom to produce a 21 cm quantum of radiation is only once in 11 million years. Despite this low transition probability, the extent of interstellar space is so vast that the total amount of emission is quite sufficient to be observed in our own Galaxy and neighbouring galaxies.

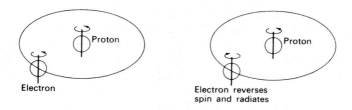

FIG. 2.15.   Production of 21 cm hydrogen line.

We have talked so far mostly about the emission of line radiation; we can also look for lines in absorption, since a good emitter must also be a good absorber. Suppose we look at a strong radio source radiating at all wavelengths. The rays on their journey to us have to traverse regions of interstellar hydrogen. A fraction of the radiation will be absorbed by giving up energy to jerk atoms into the higher energy state. Hence radiation will be absorbed at 21 cm wavelength corresponding to the hyperfine transition of atomic hydrogen. The spectrum of the distant source will show an absorption line at this wavelength.

Why is it so important to study the 21 cm radio line? One reason is that hydrogen is so widely distributed throughout our Galaxy, and in other

galaxies, that the radio hydrogen line provides a great deal of information about galactic structure. There is no corresponding optical line, so the radio data is unique. The other special feature of line radiation is that it enables velocities to be found by the Doppler shifts in frequency.

Interstellar Molecular Masers

Radio spectroscopy has led to a breakthrough in the discovery of a quite unexpected abundance of different types of molecules in the Galaxy. Optical methods have been hampered both by obscuration and by the paucity of molecular lines within the visible band. On the other hand, molecules offer a wide range of transitions of rotational energy that fall in the radio band. The radio-astronomical observations of molecular lines are discussed in Chapter 8. An extraordinary result emerging from the observations has been the discovery of interstellar masers. We shall here explain the principles of maser action.

The concept of stimulated emission was first propounded by Einstein early this century. The first practical application was devised in 1955 by Townes and his colleagues at Columbia University, USA, when they constructed a laboratory maser (the term stands for Microwave Amplification of Stimulated Emission of Radiation). To understand maser action let E_1 and E_2 be two energy states of an atom or molecule. When a substance is in thermal equilibrium there are naturally fewer atoms at the higher energy level E_2. Let us suppose that by some means the population at this higher level E_2 is increased. Then if the substance is irradiated at a frequency f corresponding to $E_2-E_1 = hf$ the effect is to stimulate transitions from the higher state to the lower state thus amplifying the incident radiation. We may liken the action to boulders perched on a slope requiring only a light push to start them avalanching down. A characteristic of spectral lines amplified by stimulated emission is their extremely high intensity and very narrow bandwidth. Of course, the essence of a maser is to devise a method to "pump" the particles to the higher level. Hence the surprise to find maser action in interstellar molecules. These interstellar masers are described later in Chapter 8.

Radiant Energy

A few points should be noted to clarify ideas on thermal and spectral line energy. According to the kinetic theory the basic thermal energy of a gas resides in the total energy of random motion of the constituent particles. It is not surprising therefore to find there is a relation between average kinetic energy of particles and the temperature given by $mV^2 = kT$. Here m is the particle mass, V the mean velocity components, T the temperature, and k is Boltzmann's constant. The German theoretician Boltzmann,

towards the end of the last century, calculated the probable distribution of particle velocities. The condition of highest probability is often described as maximum entropy. It is significant that the radiated photon energy hf at the maximum of the perfect radiator spectrum also has a value of the order of kT.

It is useful to note the magnitude of photon energy in the various radiation bands. The photon energy E is related to frequency f by $E = hf$ where h is Planck's constant. Expressing E in electron volts the relation in terms of the wavelength λ in μm becomes $E = 1.24/\lambda$. The values in different parts of the spectrum are as follows:

	Typical wavelength (λ)		Approx. photon energy (eV)	
Metre wave radio	1	m	1	μeV
Microwave radio	1	mm	1	meV
Far infrared	100	μm	10	meV
Near infrared	10	μm	0.1	eV
Light	5	μm	0.2	eV
Ultraviolet	0.1	μm	10	eV
Soft X-rays	1	nm	100	eV
Hard X-rays	0.1	nm	1	keV

It is interesting to compare the energy of electrons generating synchrotron emission with the energy of the radio photons they emit. Amazingly high electron energy is involved, of the order of 1000 MeV. It is like using a massive power station to light a match!

Shock Waves

Before leaving this discussion of the basic principles of radio we must briefly describe several kinds of waves in gaseous media that can have an influence on radio emission. For example, a shock wave accompanies a large explosive event such as a solar eruption or a supernova. Its passage through the surrounding gas raises the temperature and pressure.

The formation of a shock wave can be explained as follows. An explosion suddenly initiates powerful compression waves. The first impulse travels at the speed of sound, and the compression heats the gas. The rise in temperature increases the speed of sound, so that following impulses travel faster and catch up the initial wave. A rapid succession of compression waves thus builds up a discontinuity, a shock wave, travelling much faster than the local speed of sound and accompanied by an abrupt rise of

temperature and pressure. For instance, a shock wave with Mach number 8 (that is 8 times the speed of sound) in a monatomic gas gives a 20-fold increase of temperature.

Plasma Waves

The motion of a wave disturbance in an ionised gas is controlled not simply by collisions between particles but also by the electric forces between the charged ions and electrons. One type of plasma wave depends on the electron movements. As electrons are light particles their thermal velocity is very high (over 5000 km/sec at a temperature of a million degrees in the solar corona) and the speed of wave propagation correspondingly fast. In appropriate circumstances the plasma waves can be transformed to radiate as radio waves.

If magnetic fields are present they can exert a dominant control on the dynamic behaviour of ionised gas. An important type of wave is the hydromagnetic wave, sometimes called the Alfvén wave since it was first recognised by the Swedish physicist Alfvén. In a highly conducting plasma the magnetic lines of force are said to be "frozen" because the field lines and ionised gas move as if tied together. Induced forces prevent any relative movement. But ionised particles can move freely along the lines.

Although the field lines have no real existence except as a theoretical representation of the magnetic field they remain a valid concept in describing behaviour. From this viewpoint Alfvén waves can be pictured as vibrations of field lines. The velocity of Alfvén waves depends on field strength and gas density and is usually around a few hundred km/sec. A sufficiently strong impulse creates a "hydromagnetic shock wave". This can be a "collisionless shock", since the magnetic field can transmit impulses with the plasma frozen to it and hence is not dependent on particle collisions. These ideas serve as a guide to the admittedly complex wave phenomena in an ionised gas, particularly in the presence of a magnetic field.

The Doppler Effect

The Doppler effect is well known in sound, for example, a train whistle gives a higher note as the train approaches and falls in pitch as the train passes. This is easily explained by the compressing together of the sound waves as the train approaches, so shortening the wavelength and hence increasing the number of waves reaching us per second. As the train recedes the waves are stretched out and the frequency falls. Exactly the same process acts on light waves and on radio waves.

Consider where the waves are after 1 second. If the wave velocity is C, this means that the waves travel a distance C in a second. But in this time

the source has moved a distance equal to its own speed, V. So the fractional change of wavelength is V/C. Hence if we express the Doppler shift as a fractional change of wavelength (usually written $\delta\lambda/\lambda$), then

$$\text{Doppler shift} = \frac{V}{C}.$$

Hence by measuring the Doppler shift we can find the velocity V.

The Doppler effect can readily be observed in the 21 cm hydrogen line. If the hydrogen is moving towards us, the 21 cm wavelength becomes slightly shorter; if the hydrogen is moving away the wavelength is increased. Hence we can deduce the motion of the hydrogen and this leads to important knowledge about the spiral rotation of the Galaxy.

The Doppler shifts of atomic and molecular lines are valuable indicators of movements of gaseous regions. In addition the widths of the lines reveal internal motion due to random shifts resulting both from the thermal velocities of constituent particles and large-scale turbulence.

3. Radio Telescopes

Parabolic Reflectors

Parabolic mirrors are commonly used in optical astronomical telescopes. The U.S.A.'s largest optical telescope at Mt. Palomar has a parabolic reflector 5 m in diameter. Similarly, a parabolic reflector can be used to collect radio waves and bring them to the focus, where they are then passed on to the receiver.

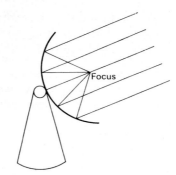

FIG. 3.1.   Parabolic reflector radio telescope.

There are of course some definite differences between optical and radio telescopes. For instance, the surface of a radio reflector does not need to be a polished mirror. It is only necessary that it is a metal, such as steel or aluminium, and it makes no difference if there is a thin layer of protecting paint over it as the radio waves go through the paint. The reflecting surface can even be of mesh, since if the holes in the mesh are small compared with the wavelength the radio waves scarcely know that the holes are there. Then another difference is in the arrangement at the focus. In the optical instrument we can look at the image or take a photograph. In the radio telescope, the radio waves are either collected in a metal horn and passed to the radio receiver down a metal tube called a waveguide, or they are picked up on a dipole where the electric field of the radio wave sets up an oscillating voltage which is then passed along wires to the radio receiver. Very often, the connection to the receiver is made by a concentric cable, consisting of an inner conductor surrounded by an outer cylindrical cover acting as the other conductor.

30

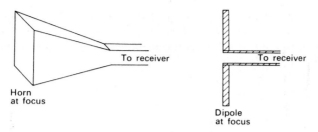

Horn
at focus

Dipole
at focus

FIG. 3.2.   Collectors at the focus.

A good radio telescope has a narrow beamwidth, and we will now explain this important concept. A radio telescope receives the maximum signal when pointing straight towards a radio source. If it points slightly away from the source it still receives some signal, and we say the source is within the beam of the radio telescope. We would like the signal to fall to zero very quickly as we move away from a source, since otherwise we cannot distinguish two sources close together in direction. What we want is a narrow beam to give good resolving power. As we shall explain in a moment, in order to produce a narrow beam the radio telescope must be very large compared with the wavelength.

Consider again the radio telescope pointing straight towards a source. The radio waves all arrive in step at the front aperture of the radio telescope as shown in Fig. 3.3.

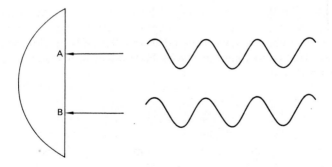

FIG. 3.3.   Waves arrive in step.

After striking the reflector all the waves arrive at the focus in the same phase of the wave motion and add together. But the waves from an inclined direction are received over different parts of the aperture slightly out of step. The combined signal at the focus therefore diminishes. The

signal falls to zero when the waves received on one half of the aperture are completely out of phase with those received over the other half. This means that crests of the waves over one half coincide with troughs over the other as shown in Fig. 3.4.

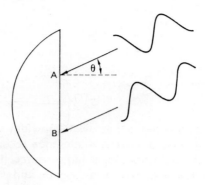

FIG. 3.4.   Waves arrive out of step.

If λ is the wavelength, and D the diameter of the aperture, then the received signal is zero when the path difference is $\frac{1}{2}\lambda$ at points $\frac{1}{2}D$ apart, as shown at points A and B in Fig. 3.4. This occurs when the angle

θ is $\dfrac{\frac{1}{2}\lambda}{\frac{1}{2}D}$ $=$ $\dfrac{\lambda}{D}$ rad $=$ $\dfrac{60\lambda}{D}$ deg approximately.

A diagram like that shown in Fig. 3.5 to represent the beam of a radio telescope is often called the polar diagram. For a receiver, the diagram shows how the sensitivity to received power depends on direction. For a transmitter, the same diagram shows how the power transmitted varies with direction.

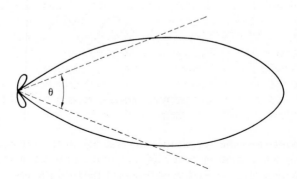

FIG. 3.5.   Polar diagram.

The angle between the directions corresponding to half the maximum sensitivity is called the beamwidth. The angle is approximately equal to the value of θ derived above, namely

$$\theta = \frac{60\lambda}{D}$$

degrees. For example, a radio telescope of 25 m diameter used at a wavelength of a metre has a beamwidth of about 2½ deg.

Terms frequently used to describe the performance of radio telescopes are the "effective area" and the "gain". The effective area is the collecting area for radio power and is normally rather less than the geometrical area of the aperture. The reason is that the feed at the focus collects the radiation better from the centre of the parabolic reflector than from the edges, so it does not receive uniformly over the whole aperture. A very good parabolic reflector has an effective area of about 60 per cent of the aperture area. The gain tells us the increase in power we receive in the beam as compared with an imaginary aerial having the same sensitivity in all directions.

The effective area (A) and gain (G) are related to each other quite simply. The beamwidth θ is λ/D rad, and so the radiation is effectively concentrated into a solid angle $\omega = \lambda^2/A$ sq. rad. Compared with a spherical polar diagram, equal in all directions covering a solid angle of 4π, the gain is $4\pi/\omega$. Substituting $\omega = \lambda^2/A$, we have $G = 4\pi A/\lambda^2$.

We can now see that there are two good reasons why we want a large radio telescope. Firstly it collects more radiation. The bigger the area of the aperture the more power is received. Secondly, it has a narrower beam and is therefore better able to resolve sources that are close together or to find the precise shape or structure of a source.

We notice also that the polar diagram in Fig. 3.5 has small lobes outside the main beam. The sidelobes are due to incomplete cancellation of waves over the aperture in these directions. They can be a nuisance when we are trying to observe the radiation from a weak source in the main beam if it happens that a strongly emitting source like the Sun is being picked up in a sidelobe. Even the radio emission from the ground picked up in the sidelobes adds unwanted noise which may be detected with very sensitive receivers. It follows that the sidelobes should be kept as small as possible, and various precautions are taken in designing aerials to minimise the sidelobes.

The main engineering problem in constructing a radio telescope is not just that of making a steerable reflecting bowl of large size. The surface of the reflector must be accurately parabolic in shape. Figure 3.6 illustrates the effect of surface irregularities.

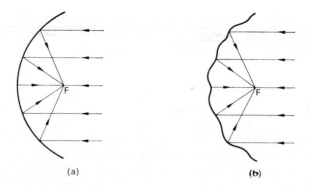

(a)   **(b)**

FIG. 3.6. (a) Waves arrive at focus in step. (b) Different path lengths to focus; waves not in step.

If the surface deviates more than a fraction of a wavelength, the waves arrive at the focus out of step and the signals no longer add together. The bigger a radio telescope the harder it is to maintain the required accuracy. Usually we specify that to work efficiently at a given wavelength, λ, the surface must not depart by more than $\lambda/16$ from the true parabolic shape. Suppose, for example, we wish to make observations at a rather short radio wavelength, say 3 cm. Then 1/16 of this is about 2 mm. It is asking a great deal of a large steerable reflector, 30 metres or more in diameter to keep all surface deviations to within 2 mm. Distortion occurs due to the sagging of the reflector and its supporting structure by its own weight, and to deformations caused by wind, and by variations of temperature producing differential expansion (especially when part of the radio telescope is in sunlight and part in shadow). All these deviations are obviously greater the larger the diameter, and it becomes increasingly difficult to maintain the parabolic profile of the reflector accurate to 1/16 part of the wavelength. Attempts can be made to counteract the deviations by strengthening the reflector frame and mount; but whatever is done the deformations increase with size, and a limit is reached to what is mechanically feasible. Along with the structural problems the cost mounts rapidly. A notable innovation in design was the 25 m steerable parabolic mesh reflector constructed in the late 1940s at Dwingeloo in the Netherlands consisting of several hundred plane facets adjusted in position after mounting. Some examples of steerable parabolic reflectors are shown in Fig. 3.7.

For many years the Jodrell Bank 76 m radio telescope had the largest fully steerable reflector, but it has now been overtaken in size and accuracy by the German 100 m radio telescope. In Australia, the construction of the 64 m Parkes telescope was eased by limiting steering in elevation to angles above 30°. The beamwidth at 10 cm wavelength is about 7½ min of arc. For

Fig. 3.7. (a) The forerunner of modern telescopes was built by Grote Reber at Wheaton, Illinois, USA in 1937. Reber is here standing beside his 9.5 m diameter parabolic reflector after its subsequent transference to the National Radio Astronomy Observatory.

observations at short wavelengths down to several millimetres a fully steerable telescope was completed at Serpukhov in USSR in 1958, and a similar telescope was later constructed at the Crimea Astrophysical Observatory. The beamwidth is about 2′ at 8 mm wavelength. For even shorter wavelengths a reflector 11 m in diameter has been installed at Kitt Peak, USA, a high site chosen to minimise atmospheric attenuation.

FIG. 3.7. (b) The CSIRO 64 m radio telescope at Parkes, Australia ($\lambda_m \sim 10$ cm).

Note: λ_m is the minimum wavelength for which the radio telescopes were designed to provide maximum efficiency. Most of the telescopes can be used effectively at shorter wavelengths than λ_m.

We have already mentioned that there is a practical limit to the size of a fully steerable radio telescope determined by the cost and by mechanical problems of achieving the required accuracy. The structural limits cannot be specified precisely, and construction of large steerable radio telescopes

FIG. 3.7.(c)   The Jodrell Bank 76 m radio telescope. The first giant fully steerable parabolic reflector.

presents a challenge to engineering skill and ingenuity. In terms of our present ideas of mechanical feasibility it would appear possible to construct say a 50 m steerable radio telescope to work down to wavelengths of a few millimetres wih about 20″ beamwidth, or a 150 m radio telescope at 21 cm wavelength giving a beamwidth of about 5′, or a 300 m radio telescope at 1 m wavelength with a 12′ beam. The cost of any of these would certainly be extremely high, especially when we bear in mind the required pointing accuracy which should not exceed 10 per cent of the beamwidth.

Meanwhile improvements in existing telescopes continue to be introduced. For instance, more accurate surface panels have been installed in the central portion of the Parkes reflector for observations at wavelengths of ~ 1 cm. Two notable radio telescopes of 45 m diameter, at Green Bank, NRAO, in USA, and at the Algonquin Radio Observatory, NRC, in Canada, originally designed for operation at several cm have improved their range to shorter wavelengths. Several research centres are developing new radio telescopes for the millimetric band.

FIG. 3.7. (d) The 22 m parabolic reflector at the Lebedev Physical Institute, Serpukhov, near Moscow ($\lambda_m \sim 8$ mm).

Reflector with Partial Steering

Let us consider for a moment what resolving power we need in radio astronomy. The Sun and the Moon both have an angular size of about 30 min of arc. Andromeda is the nearest spiral galaxy outside our own and subtends a few degrees. To distinguish features in these sources the beamwidth must not exceed a few minutes of arc. But these are comparatively large objects in the sky. If we are to study the myriads of smaller astronomical objects we must clearly improve on the resolution afforded by the fully steerable, single reflector type of radio telescope.

One way of alleviating both the high cost and the engineering difficulties of large radio telescopes is to restrict steering to elevation only, while utilising the Earth's rotation to scan across the sky. A meridian transit telescope operates in this way; it is set at any chosen elevation looking South, and objects in the sky drift through the beam. Such a method has the restriction that an astronomical source can be examined only once per

FIG. 3.7. (e) The 100 m radio telescope of Effelsburg, Germany ($\lambda_{min} \sim 2$ cm)

day as it crosses the meridian. A radio telescope of this type, 100 m in diameter, has been used at the U.S. National Radio Astronomy Observatory at Greenbank, West Virginia. Another approach is to depart from the circular aperture and have a long, low radio telescope built near the ground, so easing the structural problems. The beam shown in Fig. 3.8 is fan-shaped, narrow in a plane through the long dimension but wide in the vertical plane.

Fig. 3.7. (f) Composite picture of the Haystack installation at the MIT Lincoln Laboratory, USA. The 36 m reflector is illustrated through a "cut-away" in the radome—a plastic cover supported on a framework providing protection from wind and weather ($\lambda_m \sim 3$ cm).

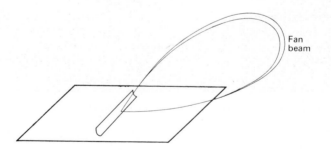

Fan
beam

Fig. 3.8.   Long radio telescope near the ground gives a fan beam.

Suppose a fan beam is being used to examine a source like the Sun. The beam at any instant receives radiation from a strip across the Sun, where the beam intersects the solar disk as shown in Fig. 3.9.

As the Sun moves across the sky the beam drifts across the solar disk. A beam like this, narrow in one direction, can be very useful in examining how much radio emission comes from different parts of a source.

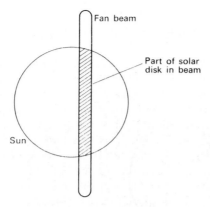

FIG. 3.9. Fan beam scans the Sun.

An interesting version of a transit radio telescope, built by Kraus of Ohio University, is illustrated in Fig. 3.10. The fan beam is tilted in elevation by means of a long flat reflector which reflects the radiation on to a long fixed parabolic reflector and thence to the focus.

A larger version of this type of radio telescope built at Nancay in France has an aperture 220 m long and 40 m high.

A giant radio telescope for operation at short wavelengths down to 8 mm and built near the ground to ensure stability has been constructed in the Northern Caucasus, USSR. The instrument known as RATAN–600 (the initials indicating a radio telescope of 600 m aperture) consists of 895 tiltable panels sited in a ring of 600 m diameter able to reflect radio waves to centrally placed receiving elements. Several modes of operation are possible. Sectors can be used for observations in different directions

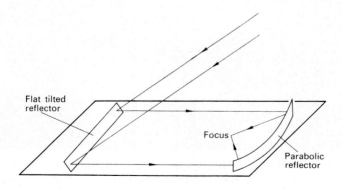

FIG. 3.10. Kraus radio telescope

simultaneously. Also, the incorporation of a secondary flat reflector enables the system to operate in the same manner as the Kraus radio telescope.

There are a few instances of reflectors built into natural hollows in the ground, thus easing the constructional problems. The best known example of this type is the 300 m dish at Arecibo, Puerto Rico, shown in Fig. 3.11. The beam direction can be altered by as much as 20° away from the vertical by displacing the feed position. Such a movement of the feed would not be permissible in a parabolic reflector as it would distort the beam. To get over this difficulty the Arecibo dish has been made spherical. This introduces a new problem for although the feed can be moved to alter the direction, a spherical surface is not the required shape for a perfect reflector. To counter this, a special feed has been designed which corrects for the deviation of the spherical reflector from the correct parabolic profile. So the ability to alter the beam direction is satisfactorily achieved by a rather roundabout process.

Fig. 3.11.   The Arecibo radio observatory operated by Cornell University, USA, has a 300 m diameter spherical reflector constructed in a natural hollow in the ground. The photograph shows the reflector, and the movable feed supported 145 m above the ground can aim the beam in any direction within 20° of the vertical ($\lambda m \sim 10$ cm).

We have now seen how it is possible to make bigger single radio telescopes by building them near the ground and by limiting steering. In this way sensitivity is increased because the collecting area is greater, as well as improving angular resolution. Until 1973 the Arecibo dish was mainly used at $\lambda = 70$ cm giving a beamwidth of 9'. Recently the reflector surface, originally of wire mesh, has been replaced by 38,400 adjustable perforated aluminium panels to allow operation at $\lambda \sim 10$ cm. An important addition for radar studies of the Moon and planets is a transmitter with 450 kW mean power at 12.6 cm wavelength. The new French radio telescope with its length of 220 m can work down to 21 cm wavelength giving a beamwidth of 5'. Now many distant radio sources have angular dimensions less than 1'. In fact for many purposes we wish to have a resolution of 1" or less to match that of optical instruments. For instance, if we are attempting to identify a radio source with a faint optical object in a sky crowded with stars, nebulae, and galaxies, we need to know the position of the radio source to an accuracy of about 1". Now at a wavelength of say 20 cm a single radio telescope giving a beamwidth of 10" would be about 5 km in diameter! The solution to this dilemma we discuss later in the section on interferometers.

Large Arrays

A comparatively inexpensive type of aerial sometimes used in radio astronomy is an array of dipoles with a flat mesh reflector behind them. The dipoles are all connected together, and the combined signal is then fed to the receiver. In the simplest form of the system, the dipoles are placed above horizontal ground to provide an upward looking beam.

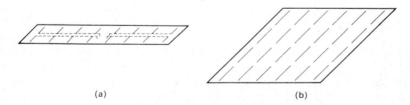

(a)   (b)

Fig. 3.12. (a) Long array of dipoles giving fan beam. (b) Square array of dipoles giving narrow beam.

The beam can be deflected by introducing phase differences between the dipole elements by altering the lengths of the interconnecting feeders. Although a rather complicated process, it is possible to arrange a set of switches to introduce different lengths of cable in steps. An array is only suitable for use at a given wavelength, because to change wavelength all

the dipole and feeder lengths would have to be altered. The advantage of this type of radio telescope is that, provided limited steering and a fixed wavelength are acceptable, a very large array can be erected. For example, a horizontal array constructed by the American scientist, Bowles, for radar studies of the upper atmosphere, covers 22 acres of ground.

The dipole with reflector is the simplest element for an aerial array, but there are other types giving more gain per element. Examples are, the dipole with a corner reflector, and a type of aerial with "directors" in front of the dipole devised by a Japanese electrical engineer, Yagi. Another type is the helix, which is a circularly polarised element. Both the corner reflector and helix aerials were invented by Kraus.

Dipole with
corner reflector   Yagi   Helix

FIG. 3.13.   Types of aerial.

So far we have discussed radio telescopes which comprise a single effective area, whether it is the aperture of a parabolic mirror, or the total area of an array. We shall next examine systems made up of a number of separate aerial units.

The Interferometer

We now consider a different approach to the problem of obtaining narrow beams and high resolution. The possibility of applying interferometer methods to optical astronomy was first suggested by Fizeau in France a hundred years ago. The initial experiments were made by Stephan, but astronomical interferometry is usually attributed to Michelson, who in 1891 successfully used the method to determine the sizes of the moons of Jupiter; later, in 1920 with the aid of Pease at the Mt. Wilson Observatory, USA, the first measurement of the diameter of a star was achieved. The introduction of interferometer methods to radio astronomy in 1948 by the Cambridge and Australian groups soon demonstrated their great value in improving radio resolution. The radio interferometer consists of two separated aerials connected together as shown, the combined signal then being fed to the receiver.

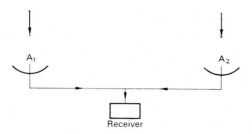

FIG. 3.14.   Two-aerial interferometer.

The line joining the two aerials is called the baseline. To understand the operation of the interferometer, imagine we are observing a distant radio source, and let us see how the signal at the receiver depends on the direction of the source. If the source direction is at right angles to the interferometer baseline, the signals at the two aerials are in step and therefore add together as shown in Fig. 3.15(a).

A short time later the source will appear to have altered its direction relative to the interferometer baseline owing to the rotation of the Earth. Consequently, the waves at the two aerials no longer arrive in step. When the path difference is half a wavelength, as in Fig. 3.15(b), the signals at the two aerials are completely out of step and cancel each other. The crest of the wave picked up at one aerial arrives at the receiver along with the trough of the wave from the other signal, and the result is zero.

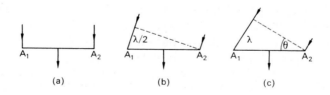

(a)   (b)   (c)

FIG. 3.15.   Combining the waves at interferometer aerials A_1 and A_2.

As the source moves along, the waves come into step again. Maximum signal is again obtained when the path difference is a whole wavelength as shown in Fig. 3.15(c). As the source continues to move, the waves alternately come in and out of phase, giving a series of maxima and minima. Referring to Fig. 3.15(c) we see that if L is the length of the baseline, then a change θ in direction of the source gives a path difference $L\theta$. Hence the signal alters from one maximum to the next when $L\theta = \lambda$, or we may write $\theta = \lambda/L$ rad ($60\lambda/L$ deg approximately).

The interferometer reception pattern therefore passes through a series of maxima and minima, the width of each lobe being $\theta = \lambda/L$. As shown on page 32, a single aerial has a beamwidth of λ/D where D is the diameter of the aperture. The effect of connecting the two aerials as an interferometer is to divide the beam of the single aerial into multiple beams, or lobes, as shown in Fig. 3.16.

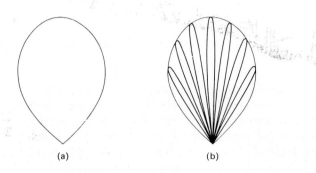

(a)   (b)

FIG. 3.16. (a) Pattern of one aerial. (b) Interferometer pattern.

The good resolving power results from the narrow width of the lobes. As the lobe width depends on the spacing L between the aerials, we can make this spacing large, for example, we could have the two aerials 2 km apart. We could hardly imagine building a single steerable radio telescope with a diameter of this size.

The receiver output of the interferometer is a series of maxima and minima known as interference fringes as shown in Fig. 3.17.

The diagram represents the output due to a point source as it passes through the field of view of the aerials. The direction of the source can be found very accurately from the pattern of fringes. At the central maximum,

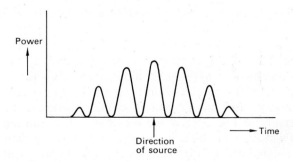

FIG. 3.17.   Received power from point source.

the source distribution is exactly at right angles to the interferometer baseline. We can also use the interferometer to measure the angular size of a source. If we are observing a point source the minima fall to zero. But if the source has appreciable extent we may regard each part as a point source and find the combined effect as in Fig. 3.18. The resultant interference pattern no longer shows zero minima.

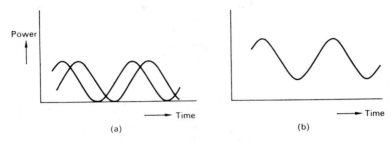

FIG. 3.18. (a) Received power from different part of source. (b) Resultant power from source.

The ratio $(P_{max} - P_{min})/(P_{max} + P_{min})$ is called the fringe visibility, and its value indicates the angular width of the source.

The Variable-spacing Interferometer

We can derive more detail about the structure of a radiating source by measuring the fringe visibility at many different spacings of the interfero-meter aerials. The process involves Fourier synthesis, named after the French mathematician who proved that any graph or map representing a structure can be formed by adding together many different sine waves of various amplitudes and phases. Observing a source with an interferometer at different lengths of baseline is a way of finding these sine waves. Combining them together enables us to derive the structure of the source in just as much detail as if we had a single aerial of the same aperture dimensions as the widest spacing used in the interferometer. We can illustrate the equivalent aperture in the following way.

When the interferometer spacing is varied along one line only, the equivalent aperture is long and narrow as shown in Fig. 3.19. Such an

FIG. 3.19. (a) Interferometer with variable spacing. (b) Single aperture.

aperture gives a fan beam; if we want a "pencil" beam, we must have a wide aperture in both length and width. An interferometer with variable spacing in two directions provides the answer. A good method is to have a variable-spacing interferometer in the form of a T as shown in Fig. 3.20. We can then separate the aerials in any direction and at any spacing up to a maximum permitted by the lengths of the tracks.

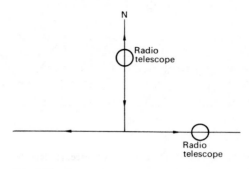

FIG. 3.20.   Interferometer with variable baseline.

In this system, one track is usually arranged to run E–W and the other N–S. To find accurate positions of sources in the sky, the E–W spacings can be used in measuring Right ascension and the N–S to measure Declination. By varying the baseline between the aerials the structures of sources can be examined.

Actually, an interferometer is more flexible with respect to its baseline than has been suggested so far. Provided the two telescopes are steerable we are not limited to looking merely at right angles to the baseline on the ground. Suppose we observe at an entirely different angle as shown in Fig. 3.21; then the effective baseline is BC.

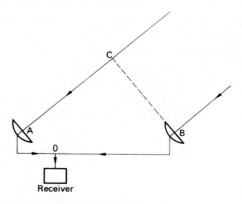

FIG. 3.21.   Effective baseline (BC) of steerable interferometer.

With steerable aerials we therefore have much freedom in varying the effective baseline. Note that the wave at *C* reaches the receiver by the path *CA + AO*, while that at *B* travels along *BO*. To equalise these path lengths, the cable length to the junction *O* must be adjusted accordingly or a compensating delay introduced.

It is worth mentioning here that a radio receiver accepts signals over a band of frequencies centred about the mean frequency *f*. Consequently, a range of slightly different frequencies (and corresponding wavelengths) are being received at the same time. The interferometer lobes coincide when the total path difference to the junction is zero. If the path lengths are unequal, the lobes no longer exactly coincide at the different wavelengths, so the fringes become blurred if the receiver bandwidth is wide; consequently equalisation of path lengths is important in planning interferometer systems.

Aperture Synthesis

The simulation of a large aperture by means of a variable-spacing interferometer can be regarded in a different way as follows. Suppose we imagine a large aperture divided up as shown in Fig. 3.22, and that we have a radio telescope just the size of one small section. If we could place it in the position of each section in turn and combine the signals together we must get the same result as for the whole large aperture. This process is called aperture synthesis. It sounds easy, but in combining the contributions of the different sections we must know how the wave motions are related to each other in phase. The best way of including the phase relation between the sections is to obtain the output of *two* sections together, in other words, with an interferometer. For instance, sections *A* and *B* in the diagram may

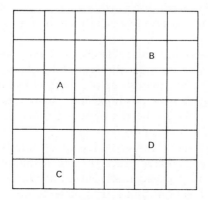

FIG. 3.22.   Synthesis of an aperture.

be taken together. The aerials at C and D would give just the same answer as A and B because the *relative* spacing is the same for both. This simplifies the procedure. We only require to find what relative positions occur in the aperture, make the observations with a corresponding interferometer representing the two elements, and then combine the outputs together in the right way to obtain the result for the whole aperture. An interferometer with aerials movable along tracks in the form of a T is clearly suitable for this type of synthesis.

Various kinds of multiple aerial systems have been devised to simulate the resolution of a large aperture. Generally you will find the clue to understanding them is that they include all the relative spacings to be found in a large aperture. If this condition is satisfied, then combining the signals in the right way will give the same resolution as a single large aperture. As we are only observing signals on small aerials, the collecting area and hence the sensitivity is much less than if we had a single radio telescope the size of the large aperture. For many investigations, however, resolving power is the most important requirement. There are thousands of radio sources in the sky and if we are to be able to distinguish them we must have a narrow beam.

The astronomy group at Cambridge led the way in the development of aperture synthesis. In the system devised at Cambridge steerable radio telescopes form a variable spacing interferometer along an E–W line. The rotation of the Earth provides the complete range of directions for each spacing. Consider two radio telescopes pointing toward the celestial North situated at A and B along an E–W line as shown in Fig. 3.23(a). As the Earth rotates, 12 hours of observations are sufficient to cover all the positions that occur in a circular strip of diameter AB as in Fig. 3.23(b). Altering the spacing AB fills a circular strip of another diameter. With many spacings a full circular aperture can be synthesised. If the radio telescopes are set to point toward a region of the sky at lower declination, the projected baseline, while the Earth rotates, covers an elliptical strip as in Fig. 3.23(c). With different spacings an elliptical aperture may be synthesised.

By introducing additional radio telescopes to observe at more than one spacing simultaneously, the total observing time can be much reduced.

The first Cambridge system utilising this method had in fact three aerials of 18 m diameter, two in fixed positions and one mobile. As the maximum spacing is approximately a mile the system became known as the Cambridge One-Mile radio telescope.

In 1972 a new and more extensive layout of aerials was brought into commission to provide greater precision and resolution by operating at shorter wavelengths over an accurately surveyed 5 km E–W baseline. With four aerials in fixed positions and four mobile aerials along a rail track a more rapid synthesis could be achieved since 16 interferometer spacings are

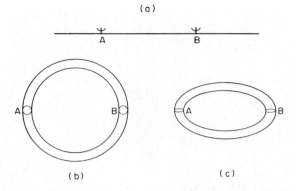

FIG. 3.23.   Illustration of the Cambridge method of aperture synthesis.

available simultaneously. The steerable reflectors, each 13 m in diameter, are designed for operation down to 3 cm wavelength with a corresponding resolution of 1″ for the synthesised beam. The primary purpose of the system is to make detailed studies of the structures of sources. In addition, by very careful surveying the radio telescope provides a precise instrument for astrometry with the ability to locate positions in the sky to 0″.1.

In 1970 the Netherlands Foundation for Radio Astronomy inaugurated at Westerbork an aperture synthesis radio telescope system consisting of a row of 10 aerials in fixed positions and two mobile aerials along an E–W baseline with a maximum spacing of 1.6 km. The steerable reflectors are 25 m in diameter and the system is planned for the rapid synthesis of the structures of radio sources at wavelengths from 6 to 50 cm. The Cambridge and Westerbork systems have much in common. Both employ steerable dishes on polar mounts and are designed to measure the polarisation of the received signals. Both take full advantage of the available advances in recording and computing techniques for the analysis and synthesis of observational data.

The most ambitious aperture synthesis system known as the Very Large Array (VLA) has recently been constructed for operation by the National Radio Astronomy Observatory, USA, at a site in New Mexico. The location satisfies the requirements for a flat high site (over 2000 m above sea level) in a dry area so as to minimise atmospheric effects, and at an appropriate latitude, 34°N, allowing coverage of all the northern sky and south down to Declination—20°. The system comprises 27 steerable radio telescopes of 25 m diameter disposed along three radial arms in the form of a Y as shown in Fig. 3.25. Each arm is 21 km long with 25 stations where radio telescopes can be positioned.

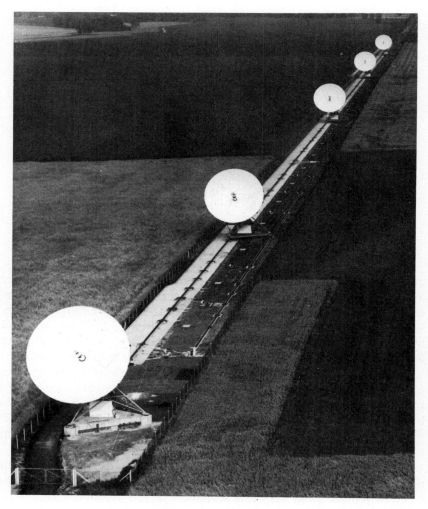

FIG. 3.24. (a) The Cambridge 5 km radio telescope showing five of the eight 13 m reflectors ($\lambda_{min} \sim 3$ cm).

Radio interconnections between the radio telescopes and the receiver in the control room are made through waveguides in the ground. The reflectors are sufficiently accurate for operation down to 1.3 cm wavelength giving a maximum angular resolution of 0″.13 for the VLA system. It is planned to use four wavelength bands up to 21 cm. The mutually connected 27 radio telescopes effectively provide 351 interferometers simultaneously. As the system has three arms, 8 hours of observation are

Fig. 3.24. (b) The Westerbork 1.6 km radio telescope comprising twelve 25 m reflectors ($\lambda_{min} \sim 6$ cm).

sufficient since the positions overlap after 120° of the Earth's rotation. This multi-linked interferometer array constitutes the best radio telescope synthesis system now available for rapid and detailed mapping of radio sources.

The Mills Cross

A different type of radio telescope is the Mills Cross, devised in Australia by B. Y. Mills. It consists of two long arrays running N–S and E–W on the

Fɪɢ. 3.25.   The Very Large Array (VLA) of the National Radio Astronomy Observatory, USA, at the site in New Mexico.

ground. The polar diagrams of the two arrays separately are two vertical fan beams intersecting each other. The region of intersection is a narrow beam, and by an ingenious method only the signals received in this narrow beam are detected. The two arrays are connected together alternately in and out of phase, and in the narrow intersection the signals alternately add and cancel. The receiver incorporates a switch-frequency rectifier so that only the switched signal in the narrow beam is detected at the output.

In the original Mills Cross, each array consisted of 250 dipoles at 3.5 m wavelength extending over a length of 500 m. The system was operated as a transit radio telescope with the beam arranged to point at different elevations towards the South. To alter the elevation of the beam, phase differences were introduced between all the elements of the N–S array by switching in appropriate lengths of cable. A new version of the Cross has been built at Molonglo, near Canberra. It has arms about 1.6 km long divided into sections with connections providing several path differences and hence several beams simultaneously. The new Mills Cross system operates at two wavelengths. At the shorter wavelength, λ = 75 cm, there are 33 separate beams of 3' width.

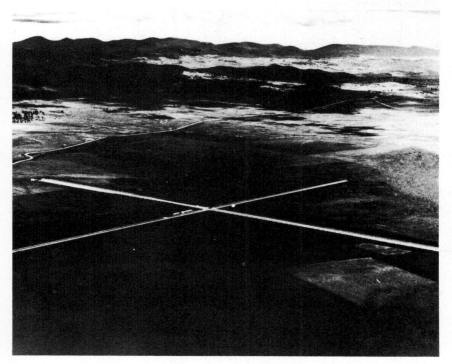

FIG. 3.26.   The Molonglo Cross radio telescope.

The Grating Interferometer

So many types of aerial system are now used as radio telescopes, it is not possible to describe them all, and only one or two further types will be mentioned here. One variety is the 'grating', consisting of a series of equal spaced receiving elements connected together as in Fig. 3.27 (a).

Suppose the radio waves from a source are incident at an angle as shown. The waves to successive elements will only add up in phase when the path differences are a whole number of wavelengths. If the path difference between the waves arriving at any two adjoining elements is $n\lambda$, then the waves add up in phase for $n = 0$ when the array faces the source, or in inclined directions such that $n = 1, 2, 3$ and so on. In consequence we have a series of narrow beams with a separation between them as shown in Fig. 3.27(b).

We can chose a spacing between the aerials so that the beams are separated by any convenient angle. Suppose we make the separation $1°$, and the grating aerial system is used to observe the Sun. Then we shall

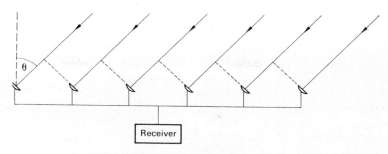

Receiver

Fig. 3.27. (a) Grating interferometer.

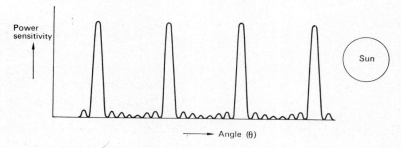

Fig. 3.27. (b) Multiple beams of grating interferometer.

receive the radio emission from the Sun as it drifts through each beam in turn. In this way the Sun can be kept under continual observation with the resolution afforded by the narrow beamwidths.

If the aerials of the grating are placed along a straight line, then a series of fan beams are formed. In this case a strip of the Sun is observed as it drifts through the beams. A better system has the aerial elements spaced along two lines at right angles connected together like a Mills Cross. The resulting beams of this system correspond to the narrow "pencil" beams where the two sets of fan beams cross each other. The first "cross-grating" was designed by Christiansen in Australia with 64 parabolic reflectors spaced along two lines at right angles, producing a group of narrow beams only 3 min of arc wide at 21 cm wavelength. As the Sun drifts successively across the matrix of beams, a map of the distribution of solar radio emission can be built up from the series of observations.

In the 1970s the system was redeveloped as an aperture synthesis radio telescope at Fleurs, Australia. The N–S and E–W arrays, each consisting of 32 equally spaced parabolic reflectors of 6 m diameter, were supplemented by two larger parabolic reflectors of 14 m diameter. Combining the outputs

of the smaller with the larger aperture elements makes it possible to suppress the grating responses (a method known as a compound interferometer originated by Covington of Canada). Although the Fleurs synthesis telescope utilises the Earth's rotation and continuous tracking to map radio sources, it differs in several respects from the Cambridge and Westerbork systems. At Fleurs there are no movable elements but more aerials. The introduction of N–S in addition to the E–W array enhances the ability to cover low declinations. The Fleurs telescope with baselines up to 800 m provided a synthesis beamwidth of 40″ at $\lambda \approx 21$ cm, and later extensions have further improved the resolution.

The Culgoora Radio Heliograph

Certain types of solar activity show large changes in very short times of the order of minutes or seconds. Bursts of radio emission, of variable intensity and position, often accompany these rapidly changing solar events. This type of radio emission predominantly occurs at metre wavelengths. What we would like to have is a complete radio picture of the Sun produced in a very short time. To achieve this aim a new aerial system designed by Wild in Australia was brought into operation at Culgoora.

The principle employed may be described as simultaneous aperture synthesis. We have explained previously the simple method of aperture synthesis for observing steady sources of radio emission by using two aerial elements first at one spacing, then at another until finally all the spacings present in a large aperture have been covered. The recordings can then be combined so as to give the resolution of the large aperture. If we now wish to speed up this process for simultaneous aperture synthesis we must have many interconnected aerials giving all the required spacings at once, and then use computers or other devices to combine the results. If different phase differences are introduced between the recordings we can synthesise narrow beams looking in many different directions at the same time. The process is obviously rather complicated, but a notable step forward to this end has been made by the construction of the Culgoora radio heliograph for obtaining high resolution pictures of the Sun at 80 MHz (3.75 m wavelength). Ninety-six steerable parabolic radio telescopes of 13 m diameter are arranged in a circle as illustrated in Fig. 3.28.

The aerials could in fact have been arranged in a cross or T formation but the circular arrangement proved a particularly convenient one. The essential requirement is that the aerials should provide all the relative spacings that are found in a completely filled aperture of the same diameter. By appropriate combinations of the aerials, it is found possible to synthesise 48 beams spaced 2′ of arc apart in N–S line, and to scan this series of beams E–W across the Sun to obtain a complete radio picture of the Sun in 1 sec.

FIG. 3.28. (a) Two views of the Culgoora radio heliograph which has 96 steerable reflectors arranged around a circle of 3 km diameter. The operating wavelength is 3.75 m.

Long Baseline Interferometers

The purpose of interferometry is twofold, firstly to find accurate directions of sources for their optical identification, and secondly to map source structures by utilising many spacings. Fulfilment of these aims demands knowledge of the phase of the recorded fringe pattern. The interfero-

FIG. 3.28. (b)

meters we have considered so far, with aerials connected at radio frequencies by cables or waveguides, meet this requirement. At spacings of a few km, source positions can be determined at cm wavelengths to better than 1″, matching that of optical telescopes and generally adequate for identifications. In mapping structure there is the practical problem of covering all spacings of the synthesised aperture without incurring excessive observational time and effort. How to counter such deficiencies, as well as the phase variations that arise in radio wave propagation will be discussed later.

Mapping distant sources of very small angular size demands extremely fine resolution. The continual pursuit of resolving powers has been an epic saga in the historical development of radio astronomy. At first, when directional accuracy better than a degree seemed a creditable achievement it was generally assumed that radio methods could never compete with optical. Now radio interferometry has forged ahead to attain a resolution of a millisecond of arc (0″.001). To realise how small this is imagine the angle subtended by a golf ball 6000 km away, roughly the distance from London to New York!

From the mid-1950s onward for more than a decade the Jodrell Bank group led the way in measurements of sources of small angular size by employing radio links to connect widely separated aerials. Initial observations at metre wavelengths with baselines ~ 1 km left a large proportion of sources unresolved and led to the deployment of longer baselines and shorter wavelengths. The progressive development culminated in an interferometer with a three-stage radio link between Jodrell Bank and Defford, a separation of 127 km, giving a baseline of 2 million wavelengths at $\lambda = 6$ cm and a resolving power of $\sim 0''.1$ in the assessment of angular sizes.

I was personally involved in this programme for as leader of a group at the Royal Radar Establishment, Malvern, we had constructed at Defford a T-shaped interferometer utilising two mobile and steerable 25 m radio telescopes. With this instrument we determined the directions of radio sources to within $1''$, the highest attainable accuracy at that time. When Sir Bernard Lovell proposed a cooperative programme of long-baseline research we readily agreed. Subsequently one Defford radio telescope became a permanent part of the Jodrell Bank interferometer network.

Long baselines and radio links bring special problems. When the interferometer aerials are widely separated the lobes become so numerous and narrow that, as the Earth rotates, the fringes pass too rapidly for normal recording. The difficulty can be remedied by incorporating a continuous phase rotator to slow down the recorded fringes. Also, in order to obtain clear fringes, path lengths have to be equalised by introducing a continuously variable delay. The radio links require a clear line of sight to operate satisfactorily, so intermediate repeater stations may be necessary over long distances. There remain the inevitable phase shifts due to atmospheric and ionospheric fluctuations. Compensation for such variations may be achieved by making rapid comparison with a nearby reference source. Finally, if practical reasons limit the aerial positions to give only partial coverage of a large aperture, the mapping procedure becomes less certain. In early measurements the aims were modest, and trial model fitting generally sufficed to derive approximate structures.

A valuable synthesis network recently developed by the Jodrell Bank group is known as the Multi Element Radio Linked Interferometer Network (MERLIN). Although it can achieve better resolution than the VLA it has limited aperture coverage. The system comprises interferometer aerials in locations partly dictated by the availability of sites and previously established links. The six sites allow 15 baselines to be formed simultaneously with separations ranging from 6 km to 133 km. Steerable radio telescopes permit tracking of a celestial region with continuously changing projected baselines. Standard microwave links convey the radio signals from the outstations to the control room at Jodrell Bank where the signals are subjected to appropriate phase rotation, delays to equalise path

lengths, correlation and computer processing. Radio maps with angular resolution between 1″ and 0″.02 are obtained depending on the wavelength. I shall describe later the methods employed to aid mapping when information is limited by partial aperture coverage and imperfect phase data.

When it became evident during the 1960s that angular discrimination of 0″.1 was still inadequate to unravel the structures of compact and distant sources it was realised that higher resolution of at least 0″.001 must be sought. To this end independent recording interferometers were devised, dispensing altogether with cable or radio links between the sites. In this way, baselines of thousands of km across continents and oceans became practicable. Two factors made the method feasible. The first was the development of atomic oscillators giving extremely stable frequency control of the local oscillators at the separated sites. The second was the advance in fast tape-recording techniques. The signals recorded on magnetic tape are subsequently played back together and processed in a computer. A Canadian group is credited with the first successful application of the method in 1967. Other groups were actively pursuing the method, and later observations have generally been organised jointly and have provided many admirable examples of international scientific cooperation. An interferometer of this type is known as a Very Long Baseline Interferometer (VLBI). A network of stations is normally planned to make simultaneous recordings. A parabolic reflector at each station continuously tracks a particular source. The range of projected baselines enables detailed source structures to be derived. The task of radio mapping when the baselines offer only partial coverage of a large aperture will now be considered.

Radio Mapping with Incomplete Data

A challenging and complex problem confronting the radio astronomer has been to devise reliable maps of radio sources from data that incompletely represents a synthesised aperture. True aperture synthesis requires that the amplitude and phase of the fringe pattern should be known for all spacings present in the simulated aperture. The ideal is scarcely ever realised in practice. The difficulty becomes particularly acute at long baselines. A map prepared from partial data is known as a hybrid map. There are two aspects to the problem. There is the undersampling due to the limited sites of the aerials; and there are phase disturbances mainly caused by atmospheric or ionospheric variations.

The coverage afforded by the interferometers is given by projected baselines at right angles to the direction of observations. A graph representing all effective spacings is called UV diagram, and a typical example is shown in Fig. 3.29. The deficiencies in coverage compared with

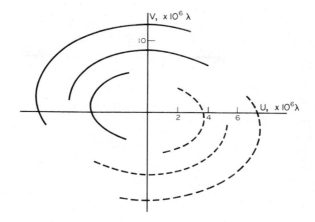

FIG. 3.29. UV diagram illustrating partial aperture coverage of a typical long baseline interferometric system. The diagram shows relative spacings in wavelengths of tracking interferometers in a plane at right angles to the direction of observation.

a complete aperture are at once apparent. There is, however, some compensation since the interferometers provide elemental units of information. An appropriate assembly of interferometer observations provides a multiplicity of equations. As a result the essential features of a radio source distribution may be derived. The requirement demands sufficient equations to define the parameters that specify the map.

Trial model fitting was the first method employed in long baseline experiments. Usually a cursory inspection of fringe visibilities gives certain guidelines to the type of structure. Once a trial model is proposed, it can be tested to see how well the calculated fringe visibilities compare with those observed. The model can then be improved until a reasonable fit is obtained. Model fitting has two defects. It is extravagant in computing effort, and the solution may not be unique.

A procedure now commonly employed in source analysis is called CLEAN. Devised by a Swedish astronomer Högbom, it has the merit of being both simple in principle and effective. A non-uniform coverage of aperture spacings corresponds to a distorted synthesised beam. The shape of the beam can be calculated. A map of the source derived from a Fourier transform of the available amplitude and phase recordings would suffer similar distortion. CLEAN is an iterative process aimed at removing from the "dirty map" the distortion due to the "dirty beam". To achieve this the beam shape is subtracted from the most prominent maximum of the radio map. In this way the highest peak and its associated distorted sidelobes are removed. The same procedure is then successively applied to the remain-

ing features on the map. Finally the map is restored by returning at appropriate amplitudes the various components with a "clean beam" (that is, the beam of a completely filled aperture).

A different method of map restoration, regarded as theoretically ideal but more difficult in practice, is known as the maximum entropy method. The term entropy originates from thermodynamics, and a statistical analysis by Boltzmann established the identity of maximum entropy and probability in physical distributions. In the radio context the maximum entropy method ensures the simplest map compatible with the available data. The map then contains the least amount of unsubstantiated information. The difficulty lies in devising the algorithm, that is, the computing procedure.

A serious problem arises in very long baseline interferometry when reliable determination of the phase of the interference pattern becomes virtually impossible. At short wavelengths irregular changes in propagation result from atmospheric variations in temperature and water vapour content. Long wavelengths are affected by ionospheric fluctuations. However, instantaneous phase of the fringes can be recorded, and by combining observations of several interferometer pairs the errors can be made to cancel. By this means reliable phase information known as "closure phase" can be extracted. This remarkable result was first demonstrated in 1958 by Jennison while engaged in research at Jodrell Bank. Surprisingly, for many years the potentialities of the method were not fully appreciated. Revived in recent years, it now represents a breakthrough in its contribution to mapping techniques.

To understand the principle, consider an interferometer system formed by three aerials. Let us denote the phase error in linking aerials A and B by ϵ_{AB}. Adding this to the error ϵ_{BC} recorded by the interferometer aerials B and C yields a total error ϵ_{AC}. But this is the same as the error in the recording between aerials A and C, so on subtraction the errors cancel. Combining observations in this manner in a closed loop gives a correct phase relation, $\phi_A + \phi_B - \phi_C$, called the closure phase, between the three aerials. The derived map of source structure must conform to the value of the closure phase. A network of interferometer aerials gives further closure-phase relations. The closure procedure can be extended to amplitudes also. Much ingenuity has been shown in the task of incorporating these values to deduce the radio maps. One way is to start with a trial model which is then adjusted to agree with the closure data; the process CLEAN is then applied as the final stage.

The success of the analytical methods described has in recent years brought a major advance in detailed radio mapping of compact and distant radio sources. With the aid of VLBI the goal of a millisecond of arc resolution has been attained. Some of the remarkable achievements in mapping radio structures are illustrated in later chapters.

Scattering by Irregularities

Irregularities in the transmission path of radio waves are a nuisance to interferometry and aperture synthesis, although they have incidentally brought a bonus of interesting information. Radio propagation from source to observer can be affected by inhomogeneities in the atmosphere, ionosphere, interplanetary medium, and interstellar space. We shall briefly outline their influence on radio observations.

In the atmosphere the variations of temperature and humidity cause changes in refractive index and distort the wavefront. The differences in equivalent path measured in wavelengths are greatest in the centimetric and millimetric bands. The relative phase at two separated aerials as in an interferometer suffer corresponding variations. For instance, departure of 30° in phase angle can occur at 5 cm wavelength at aerials separated by 1 or 2 km. The largest variations occur in summer daytime when the water vapour content is greatest. The period of received phase change at the ground depends on the wind speed and is typically about 10 minutes.

In contrast, the irregularities of electron distribution in the ionosphere influence long wavelengths most because the oscillations of electrons are strongest in electric fields of low frequency. The ionospheric inhomogeneities distort the phase, with consequent deviations in wave directions. When the refracting layer is sufficiently distant as in the case of the ionosphere, waves received together in differing phases combine to produce changes in amplitude also. It was in fact the ionospheric amplitude fluctuations that led to detection of the first discrete radio source, Cygnus A. At 5 m wavelength scintillations of 50 per cent in amplitude may occur with angular variations in direction more than 1'. Irregularities in electron content between 1 km and 1000 km exist in the ionosphere, but the most typical scale is a few km associated with the night-time sporadic F layer. Wind speeds in the ionosphere are generally of the order of 200 m/sec, and periods of fluctuation in received intensity $\sim$ 10 sec. To obviate as far as possible the effect of atmospheric and ionospheric scattering the wavelength chosen for aperture synthesis must be neither too short nor too long, and the optimum range lies between 5 cm and 50 cm.

Two other causes of scintillation are the irregular electron distribution in interplanetary space due to the solar wind, and the widespread thin clouds of ionised hydrogen that pervade interstellar space.

We shall describe later in Chapter 5 the properties of the solar corona and the formation of the solar wind. Here we shall be concerned with the influence on radio propagation of the continual outflow of plasma from the Sun extending into the solar system. Distant sources of small angular size suffer fluctuations in amplitude and angular spread through scattering in this ionised interplanetary medium. The scintillations have been put to

useful purpose in studies of angular dimensions of sources. The wavefront from a distant source undergoes distortions in plasma blobs which are more intense and smaller closer to the Sun, the scale ranging from about 10 km close to the Sun to 500 km at the distance of the Earth's orbit. The scale can be inferred by correlating fluctuations (which have periods of the order of a second) when observed at spaced receiving sites. The speed of the solar wind, about 300 km/sec, determines the rate that the pattern moves across the ground.

The magnitude of scintillation depends on the size of the source. If the source has appreciable size the displaced patterns received from different parts even out. The angular size of any given source can be deduced from the depth of scintillation compared with known very small sources used as references. As the Sun traverses the sky the solar elongation (that is, difference in direction) of a source changes. Hence the line of sight to the source covers a wide range of solar plasma scales. By observing the magnitude of scintillations we can infer source sizes between $0''.01$ and $1''$. The method has the advantage of providing very simply high resolving power at comparatively long radio wavelengths. To detect weak sources aerial arrays covering more than 2 hectares of ground have been constructed by Hewish and his team at Cambridge. In this way large numbers of sources can be surveyed at low cost.

By a remarkable stroke of serendipity the techniques developed for the study of scintillations proved exactly right for the discovery of pulsars (described in Chapter 8), and curiously enough it was the variations of pulsar amplitude that first drew attention to the occurrence of interstellar scintillations. Clouds of weakly ionised interstellar hydrogen, around 100,000 km in extent, cause fluctuations of intensity with periods of the order of minutes. The fluctuation pattern moves across the ground with a speed of 200 km/sec, roughly the velocity of motion of the solar system in the Galaxy. The magnitude of scintillation on very small sources is deep at all radio wavelengths, indicating that large phase changes have been incurred in the extensive ionised clouds. Accompanying the scintillations there is an angular spread. The irregularities have most effect near the plane of the Galaxy where the hydrogen density is greatest. A consequence is that sources exhibiting interplanetary scintillation are notably absent within about $10°$ of the galactic equator. The reason is that apparent sizes of such sources have already been much increased by interstellar scattering. Even at high galactic latitudes the scattering imposes a limit to the minimum angular diameter that can be resolved. Although the angular spread may be small it sets an important restraint on the effectiveness of VLBI at the longer wavelengths. The angular scattering exceeds $0''.1$ of a few metres wavelength; and to achieve a resolution of $0''.001$ (millisecond of arc) requires a choice of wavelength approaching 10 cm.

Lunar Occultations

Measurements during lunar occultations provide another indirect method of examining distant sources. A large aerial aperture is used to detect weak sources, and a resolution of $0''.1$ can be attained over a wide range of wavelengths. The determination of the positions and sizes of sources by observing their occultation by the Moon is simple in principle. If the Moon passes across a radio source the changes of received intensity indicate the position and structure of the source. For instance, a small source is cut off abruptly; while for a double source the received intensity falls in two steps as first one component and then the second are obscured by the Moon. At Ootacamund in India a large parabolic cylinder radio telescope at $\lambda \sim 1$ m has been installed specifically for surveys of sources by lunar occultations. The difficulty with the method is that it is applicable only to a limited band in the sky where sources happen to lie in the path of the Moon.

The Radio Receiver

The radio receiver for making observation in radio astronomy is in many respects the same type as used in radio and television sets. It has, however, to be extremely sensitive and stable. It is often called a radiometer because it is used to measure the received radio power. In radio astronomy, the "signal" we receive shows rapid and random variations of amplitude known as "noise", but we can measure quite accurately the mean level of power that we are receiving.

The most important requirements for a sensitive receiver are firstly, that the radio components should not themselves generate appreciable noise, secondly, the input should have a wide bandwidth, and thirdly, the output should be averaged over as long a time as possible. A sensitive radio astronomy receiver may detect changes of effective temperature of $0.01°C$ or less. The essential characteristics of a good receiving system are outlined in Appendix B.

4. Radio Emission from the Moon and Planets

The Planetary System

The planets, including the Earth, moving in their orbits round the Sun, constitute the principal objects of the solar system. Some of the planets have natural satellites circling round them, but most of the satellites are too small and too distant for their radio emission to be detected. The important exception is the Earth's satellite, the Moon, which subtends the largest angular size of all planetary objects because it is relatively near to us. The Moon and planets are all cool bodies (compared with stars), and depend on heating by the Sun's radiation to maintain their temperature. Before we discuss radio observations, let us get a clearer picture of the planetary system. Figure 4.1 illustrates the orbits and periods of the planets.

Most of the planets move in nearly circular orbits round the Sun. Almost all the orbits lie close to the same plane as that of the Earth. The plane of the Earth's orbit is known as the Ecliptic, a name derived from "eclipse" because it is in this plane that eclipses occur when the Earth and the Moon (or planet) are in line with the Sun. Curiously enough, the two planets

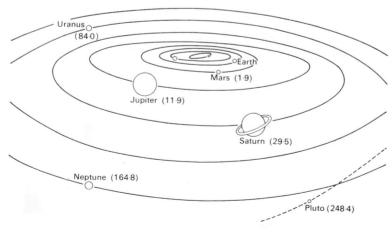

Fig. 4.1. Illustration of planetary orbits (not to scale) and periods (in years). The inner planets are Mercury (88 days) and Venus (225 days).

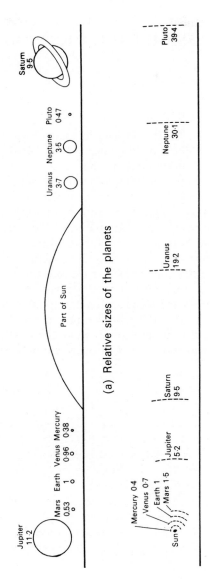

(a) Relative sizes of the planets

(b) Distances of planets from the Sun

FIG. 4.2. (a) Sizes of the planets and the Sun compared with the Earth. (b) Distances of the planets from the Sun relative to the Earth's distance.

whose orbits are most tilted out of the ecliptic plane, and are also far from circular, are the planets nearest and furthest from the Sun, namely Mercury and Pluto.

Although it is impracticable to portray the sizes and distances of the planets to scale on a single diagram, we can illustrate these parameters separately. Figure 4.2(a) shows the relative sizes of the planets, while Fig. 4.2(b) illustrates on a different scale their distances from the Sun.

The inner planets Mercury, Venus, Earth and Mars have comparable densities and are known as the terrestrial planets. They are in marked contrast with the giant outer planets, Jupiter, Saturn, Uranus and Neptune, all huge low-density, rapidly rotating objects with dense atmospheres and sometimes called the Jovian planets. Pluto is an exception—a much smaller outer planet similar in size to the Moon.

We now come to the question of the nature of the planetary surfaces, their temperatures and atmospheres. What conditions would a man encounter if he attempted to land on the planets from a space vehicle? Assuming the planets are heated solely by radiation from the Sun we can calculate roughly their average temperatures. In this way we estimate, for example, a temperature of the order of 500 K for Mercury, the closest planet to the Sun, and about 50 K for the outermost planets, Uranus, Neptune, and Pluto. Remembering that 273 K = 0°C, we see that the temperature of the planets may be expected to range from more than 200°C above to 200°C below the freezing point of water.

The best way of discovering what conditions are really like on the planets, prior to space landings, is to make observations over a wide range of wavelengths. Our knowledge of the planets comes from studies made in three different parts of the spectrum, visible, infrared and radio. Firstly, we can see and examine the planets by the sunlight reflected from them. In some cases the solid surface of the planet is visible; but in others, like Venus, there is a thick cloud layer, preventing us from seeing the surface. Although the planets look bright by the sunlight they reflect, most of the solar radiation falling on them is absorbed. The planets themselves emit thermal radiation according to the temperature they acquire. The thermal emission from the surface can best be detected at infrared wavelengths. As explained on page 13, an object at a temperature between 50 K and 500 K radiates its maximum energy at infrared wavelengths. Consequently surface temperatures are best measured by telescopes using detectors sensitive to infrared radiation. In the case of a planet like Venus, with an atmosphere opaque to infrared as well as light, the temperature measured by infrared is that of the cloud layer.

When temperatures are determined by radio emission there is an interesting difference. Radio waves can penetrate not only through cloudy atmospheres, but can also "see" through the upper crust of non-metallic solid surfaces. The temperature measured at radio wavelengths is conse-

quently an average temperature from the surface layers down to a depth of the order of a few wavelengths. Observing at longer wavelengths enables us to "look" at greater depths below the surface.

By making these observations at various radio wavelengths, and combining these with infrared and optical observations we can learn much about surface conditions and atmospheres of the planets. Before proceeding to describe the results in more detail, we must note some important practical factors that have to be taken into account. A glance at Fig. 4.1 indicates that the distances of the planets from the Earth must vary considerably as they proceed round their orbits. Let us illustrate the variation of distance from the Earth in the case of Venus.

Venus is closest to us when both the Earth and Venus are on the same side of the Sun; the distance is six times greater when Venus is on the far side of the Sun. The angular diameter is 64″ when nearest to us but only 10″ at its farthest point. Obviously, Venus is most favourable for radio observations (either by its emission or by radar echoes) when passing closest to the Earth.

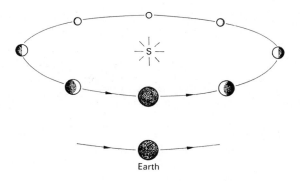

FIG. 4.3.   Orbit of Venus as seen from the Earth.

The percentage variation in distance is much less marked for the outermost planets, for their distances remain very large for all orbital positions. The planets whose orbits are fairly close to the Earth are most affected, and the corresponding changes in angular diameters at closest approach and furthest distance from the Earth are tabulated below.

A glance at the angular diameters emphasises the difficulty of detecting the outermost planets. Not only are they extremely cold, but the angular sizes they subtend are very small. Consequently, the radio astronomer is endeavouring to observe the radiation from a very low temperature object occupying only a small fraction of the beam of the radio telescope.

As well as distance, there is another factor to be taken into account in

ANGULAR DIAMETERS IN SECONDS OF ARC

	Mercury	Venus	Mars	Jupiter	Saturn	Uranus	Neptune	Pluto
Maximum	12.9	64	25.1	49.8	20.5	4.2	2.4	0.17
Minimum	4.7	9.9	3.5	30.5	14.7	3.4	2.2	0.17

considering the radiation from the planet, namely, the phase of illumina-tion. Examination of Fig. 4.3 shows that if Venus is between us and the Sun we are looking at the dark side of the planet, while in the far position, Venus is fully illuminated; between these extreme positions Venus is half bright. The planet thus goes through phases of illumination just like the Moon, although Venus is too distant for us to discern the phase directly with the unaided eye. The phase variation is large for planets with orbits close to the Sun. On the other hand, the outer planets are always seen in aspects almost fully illuminated by the Sun. The surface of a planet must clearly be hotter where the Sun is shining on it, and this must be taken into account when explaining the emission measurements. What happens below the surface, whence some of the radio emission arises, is a more complex problem depending on the thermal properties of the material, and how fast the planet is rotating. It is evident that interpreting planetary measure-ments is no easy game, but as we shall show, some fascinating conclusions have been unravelled from the clues provided by observations. Let us now look at some of the information that has been derived about the Moon and the planets.

The Moon

The Moon has a diameter approximately a quarter that of the Earth and its distance is roughly 380,000 kilometres. In terms of astronomical distances it is close to us, and consequently, subtends a fairly large angular size ($\frac{1}{2}°$). It always keeps the same face toward the Earth, apart from slight swaying known as libration. Even modest optical telescopes can pick out the main features of the lunar landscape. Early mappers mistakenly thought the vast arid plains were seas, and maritime names (in Latin) like Mare Tranquilli-tatis (Quiet Sea) and Oceanus Procellarum (Ocean of Storms) are still retained. It was subsequently realised that the Moon has a dry surface and negligible atmosphere. It has many different topographical features includ-ing mountain ranges with heights up to 8000 m. There are numerous craters consisting of a depressed area surrounded by an almost circular rim, and the largest is about 250 km across.

The appearance of the Moon in optical telescopes gave a clear lead to the natural forces influencing surface conditions. The vast mare areas look like extensive solidified lakes of lava. The huge fissures and hills could have been formed by volcanic eruptions and moonquakes in the lunar crust as the Moon cooled from its original molten state. The lunar surface is also seen to be pitted with countless craters caused by the impact of meteorites. Since there is no atmosphere to retard and burn up the meteorites, they strike the surface without loss of speed or size. This provides a process of continual erosion because in addition to large meteorites the Moon is subjected to the unceasing bombardment of tiny grains, known as micro-meteorites, pulverising the surface and ejecting a dusty spray of rock particles.

The progress of space flights has resolved many problems on the precise nature of the Moon's surface. Even so, it is interesting to see what knowledge has been derived from ground-based observations. After all, in planetary exploration remote surveillance is a necessary prerequisite of closer inspection and surface landing.

Radio emission from the Moon was first detected in 1946 at centimetre wavelengths by Dicke and Beringer in the USA, and a few years later a more complete series of measurements was made in Australia by Pidding-ton and Minnett. It was found that the phase variation of the radio emission lags behind the visible phases of the Moon. The radio tempera-ture comes to a maximum about 3½ days after the full Moon. We can understand this lag in terms of the gradual heating up of lower layers after the Sun's rays strike the lunar surface, and the gradual cooling off when the Moon's surface becomes dark. The delay is like that of a kettle of water put on an electric hot plate. The water warms up gradually, and when the heat is turned off it cools only slowly. As the region from which the radio emission originates extends to depth beneath the surface we will expect a phase lag. The magnitude of the temperature variation and the lag tells us something about the thermal capacity and conductivity of the lunar material. It soon became evident, both from infrared and radio measure-ments, that the lunar material must differ from terrestrial rocks. It was found to possess very low thermal capacity and conductivity and it was concluded that it could be something like pumice, only with more porous open structure, or it might consist of some light granular material like volcanic ash. Later measurements subsequently covered a wide range of wavelengths. At longer wavelengths (of a metre or more) the phase variation tended to disappear. This is to be expected, because at sufficient depths the surface will have completed its temperature cycle before the heat variation has had time to penetrate down. The results show that the Moon has a mean temperature of about 230 K, and superimposed on this a phase variation apparent at wavelengths less than a few centimetres as shown in Fig. 4.4. In these results, allowance has been made for the fact

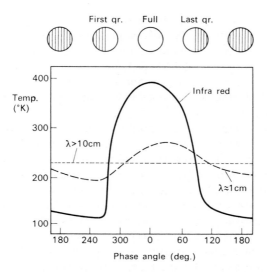

Fig. 4.4.   Temperature at the centre of the Moon's disk at radio and infrared wavelengths. (For discussion of data, see Sinton, 1962; Mayer, 1964).

that the Moon is not a perfect radiator. We know that it has a reflectivity of about 10 per cent and therefore only radiates with about 90 per cent efficiency. Hence the measured temperatures have to be increased accordingly in order to give the true temperature.

The difficult problem was to deduce with more certainty what the lunar surface must be like. Some contended that the detailed results could only be explained if there was a layer of dust above more solid ground below. Despite the limitations the radio studies have been a valuable exercise giving considerable information about the nature of the ground below the surface. For example, the Russian investigator Troitsky deduced that the density doubled in the first few cm below the surface, and that porous material extended down to a depth of about 6 m. All the clues pointed to dry dust or granular material overlying denser rock.

Early radio observations were made using fairly wide beams, in some cases containing the whole surface. When larger telescopes were used at short wavelengths giving narrower beamwidths so that the Moon's surface could be scrutinised, it became evident that different parts of the Moon's surface behaved in different ways. For example, it was found that the maria heated up more rapidly than the mountainous regions. Infrared studies also revealed many hot spots on the lunar surface, most of them being large craters. No wonder then that there were problems in interpreting the observations with wide beams believing all the Moon's surface to behave in the same way. The optical picture of the Moon, of course, makes

the heterogeneous nature of the lunar features very apparent. As Galileo wrote in 1610, "The Moon is full of inequalities".

The radio knowledge is, of course, not limited to observations of radio emission. The study of radar echoes provides additional information about the surface including average slopes and roughness; and will be discussed in Chapter 6. The results reinforce the general conclusions about the nature of the surface.

The US Apollo manned missions to the Moon and the landings of Soviet Luna spacecraft have been exciting events in lunar exploration. Many photographic and scientific recordings were made, and nearly 400 kilograms of lunar rock and surface "soil" brought back for analysis.

The results confirmed the vast areas of basaltic lava from volcanoes that erupted several thousand million years ago. The highlands are formed from upheavals of solidified rock from lower levels. Major craters are the consequence of an immense meteoric bombardment that occurred in the early stages of the Moon's history. Subsequent meteoric impacts have caused continued erosion. The rocks are dry for there is no water on the Moon, and with low gravity and no atmospheric resistance the debris of meteoric impacts spreads over wide areas. The dust and "soil" formed from meteoric particles and fragmented rocks indeed covers the lunar surface as the early radio measurements had predicted.

Venus

Venus comes closer to the Earth than any other planet. The positions and phases of Venus in its orbit are illustrated in Fig. 4.3. It is nearest to us when it comes between the Earth and the Sun, a position known as inferior conjunction. The Earth is then facing the dark side of Venus, which is consequently difficult to see at this time. As Venus moves farther away from the Earth, more of the illuminated surface becomes visible. On the other hand, the angular size of Venus becomes smaller due to its greater distance. Venus therefore has maximum optical brilliance at an intermediate position, during the crescent phase of illumination.

The surface of Venus is obscured from view by a complete coverage of white clouds that reflect light strongly, and contribute to its high visual brilliancy. Even the rate of rotation of the planet cannot be deduced optically because the clouds are so dense. The ability of radio waves to penetrate through clouds provided the opportunity to explore the surface and atmospheric conditions. Generally speaking, to measure temperatures of relatively cool objects like planets, the obvious choice of wavelength is infrared. However, as the clouds of Venus are opaque to infrared, the temperature found in this way is that of the cloud layer. The measurements indicate a mean temperature of about 225 K at inferior conjunction, with scarcely any change with the phase of solar illumination. At first sight, the

constancy of this value may seem surprising, but it must be remembered that this is simply the atmospheric temperature at a high level. Not much of the solar radiation is absorbed here, so we have no strong reason to expect the temperature to follow closely the solar illumination. Atmospheric temperatures are largely influenced by circulation and convection.

The detection of radio emission from Venus by Mayer and his colleagues in USA in 1956 gave the first direct measurement of the temperature at the surface. No one had anticipated such a high radio temperature as they recorded at 3 cm and 9 cm wavelength, namely about 600 K. Taking into account that the surface cannot be a perfect radiator the true temperature must be even higher.

Due allowance for the emissivity leads to a true temperature of more than 700 K. The dense cloudy atmosphere is more opaque at shorter wavelengths so that part of the radiation at millimetric wavelengths arises from the higher and cooler atmospheric regions. For example, at 8 mm wavelength an intermediate temperature of 400 K was recorded. The temperature difference from about 750 K at the surface to 225 K at the top of the clouds indicates a very high and dense atmosphere.

The extraordinary surface temperature revealed by radio measurements demanded explanation. Optical and infrared observations of spectral lines had yielded an important clue—that carbon dioxide (CO_2) is very prevalent in the atmosphere of Venus. It had been predicted as long ago as 1940 that the abundance of CO_2 would produce a strong "greenhouse" heating effect. For CO_2, like a glass window, has the property of trapping solar radiation. Satellite observations have confirmed the massive Venus atmosphere, with a ground pressure more than 90 times that on the Earth, and that CO_2 is the main constituent contributing to the remarkably intense "greenhouse" effect.

The Russian Venera and the US Mariner and Pioneer space probes have revealed much detailed data on the composition of the atmosphere. The topmost layer contains a haze of sulphuric acid droplets, and within the clouds below are large numbers of sulphur particles. The cloud cover reflects almost 80 per cent of the sunlight it receives, hence the brilliance of Venus in the night sky. At the surface of Venus, space probes have revealed a rock-strewn landscape in a lurid red light which is all that remains of the sunlight dimly penetrating the dense and inhospitable atmosphere. Very frequent lightning discharges have been recorded, and are thought to contribute the ashen glow on the nightside of Venus. It is curious that the atmospheres of the Earth and Venus are so different for these planets nearly equal in mass and size. This startling anomaly has led to conjecture on the dangers of terrestrial pollution and deforestation increasing the amount of CO_2 in the Earth's atmosphere. We shall return later to the important role of radar in delineating the topography of the surface of Venus.

Mercury

Mercury is the planet closest to the Sun with an orbital period of 88 days. It is also the smallest planet, with a diameter of 3200 miles. With its arid surface and sparse atmosphere, it bears a superficial resemblance to the Moon. Because it is so close to the Sun the temperature of the illuminated surface is very high, and infrared measurements showed that the temperature in full sunlight reaches 610 K. It had long been believed that the planet always kept the same face to the Sun, and that the dark side remained intensely cold at a temperature not much above absolute zero. The first radio measurements of Mercury at the University of Michigan, USA, in 1961 occasioned surprise because they indicated a mean temperature of about 250 K on the dark side, much higher than had been anticipated. Subsequent radar studies of Mercury have proved that the planet is slowly rotating, and consequently the whole surface is periodically warmed by solar radiation. This explains the relatively warm sub-surface temperature on the dark side.

The observation of Mercury at all its phases is hampered by its proximity to the Sun with its intense radio emission. Nevertheless, measurements at radio wavelengths have shown phase variations like those of the Moon. At 2 cm wavelength there is a marked variation with the phase of illumination, while at 11 cm wavelength the temperature is almost constant at all phases. As we found for the Moon, the temperature changes are greatest near the surface. The longer wavelengths "look" at the radiation from lower depths where the temperature variation is much less. Photographs taken during the US Mariner 10 spacecraft fly-by in 1974–5 revealed that Mercury has a cratered surface resembling in many respects that of the Moon.

Mars

Mars is the first planet farther out than the Earth in the planetary system. The detection of radio emission from Mars was first reported in 1958 in the USA, and the mean temperature of the illuminated surface measured at closest approach, was found to be about 200 K. Mars, with its superficial resemblance to the Earth, its channels, varied surface features, and climatic changes, has long provoked speculation and interest. Its radio emission has not, however, appeared to be a particularly attractive subject for study, and has received relatively little attention. In comparison, the US Viking spacecraft orbiting and landing on Mars have had obvious advantages, and have been a most fruitful source of photographic and other detailed data.

Jupiter

Jupiter is the most remarkable planet of the solar system. It is a colossus in size, with a mean diameter of 88,800 miles (142,700 km), over 11 times that of the Earth. Despite its great size it is spinning more rapidly than any other planet, the rotation period being less than 10 hours.

Because the planet is so big, Jupiter subtends on average a larger angular diameter than any other planet, about 40″, and is one of the most brilliant objects in the sky. Inspection through a telescope reveals a very cloudy atmosphere with a system of markings illustrated in Fig. 4.5(a). The most characteristic features are the pink or purplish-brown belts and greyish polar caps. The markings vary, suggesting a very turbulent atmosphere, but there is a notable semi-permanent feature, the Great Red Spot. Jupiter has 12 satellites circling round it; four of these are of substantial size, with diameters similar to the Moon. The principal satellites, Io, Europa, Ganymede, and Callisto, shown in Fig. 4.5(b), are known as the Galilean satellites since they were first observed by Galileo.

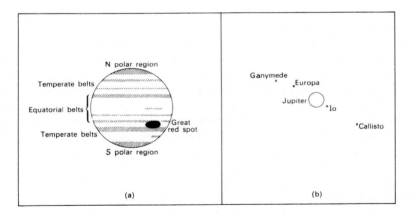

FIG. 4.5. (a) Illustration of Jupiter's belts. (b) Jupiter's principal satellites (as seen in the field of view of a small telescope).

The rate of spin of Jupiter has been deduced visually from the motion of the spots and markings in the belts. The rotation period of the atmosphere varies slightly, and equatorial regions rotate faster than the temperate regions. This has led to two systems for defining positions on Jupiter; System I for the equator, and System II for the poles. The cause of the Great Red Spot is not known; it appears to be the top of an atmospheric vortex originating from a lower level.

Although the accomplishments of the US Voyager spacecraft in 1979 resulted in a dramatic increase of information on Jupiter and its satellites, the previous ground-based radio studies had remarkably extended our understanding of the properties of the planet. I shall now review the knowledge gained by radio observations from the Earth.

Decametric Radio Emission from Jupiter

Jupiter was the first planet to be observed by radio. The discovery of intense bursts of radio emission was most remarkable and unexpected. In 1955 Burke and Franklin in Washington, USA, were testing out an aerial array for mapping radio noise sources at a wavelength of 13 m. At certain times, their records appeared to be affected by bursts of interference. When one night they noticed Jupiter lying in the direction of their aerial beam at the time the bursts occurred, they treated the coincidence as a joke, because it seemed so improbable that Jupiter could be responsible. However, a careful comparison of their records of the time of occurrence of recorded bursts with the position of Jupiter proved the planet to be the source. Only the Sun rivals Jupiter as a source of powerful outbursts at decametric wavelengths.

The next step forward was made in Australia by Shain who re-examined earlier recordings of cosmic radio noise at 17 m wavelength. He also found the Jupiter bursts although he had previously dismissed them as interference. Analysing his records to see whether the bursts originated from the Great Red Spot, or some other part of Jupiter, he found that the occurrence of the bursts varied with the rotation of the planet, with one major peak in the bursts and two lesser ones. These peaks were not associated with any visible features, and Shain noted that the radio period of rotation was different, being slightly faster than System II. He therefore designated a third system for reckoning rotation and longitudes on Jupiter known as System III. The rotation rate indicated by the radio bursts on System III is remarkably steady, with a period of 9 h 55 min 29.37 sec. The occurrence of radio bursts on System III longitudes is illustrated in Fig. 4.6.

The peaking of the bursts was interpreted as due to a particular source (or sources) sending out a directional beam of intermittent bursts of radio emission which we detect when the source is facing toward the Earth.

We will now describe some of the main properties of these unexpectedly powerful bursts of radio emission from this cold massive planet. They have been observed at wavelengths from about 8 to 50 m, but their occurrence is most frequent at about 17 m (18 MHz). They are often strongly circularly polarised, so giving a clear evidence of a magnetic field. It is probable that the wavelength of emission indicates the escape of radiation from a Jovian ionosphere with a critical frequency near 18 MHz (and therefore having an electron density 10 to 100 times greater than the terrestrial ionosphere).

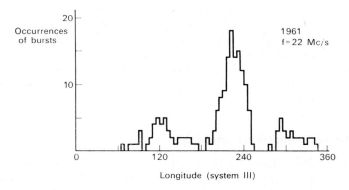

FIG. 4.6.   Dependence of Jupiter bursts on longitude. (After Douglas and Smith, 1963.)

The bending of ray paths in the Jovian ionosphere could explain the apparent beaming effect. The circular polarisation indicates that the radio frequency of the bursts is close to the gyro frequency and this leads to an estimate of the magnetic field. As explained in Chapter 2, the gyro frequency in a field of strength H is $28H$, and for a frequency of 18 MHz this yields a value of $H \approx 0.7$ mT, more than twenty times stronger than the Earth's field. There is good evidence that the bursts are less evident near sunspot maximum, and this could be due to the Jovian ionosphere then being denser, so restricting the escape of radio emission.

All this may well be correct, but of course it does not explain the origin of the emission which is still very puzzling. The radio emission is so immensely more powerful than any lightning flashes on the Earth that "thunderstorm" theories have been abandoned. The "brightness temperature", even if the whole disk of Jupiter were radiating, corresponds to over 10^{15} deg K. Actually the source size is only a tiny fraction of the disk as recent interferometer measurements with long baselines have shown. Measurements at $\lambda = 15$ m with baselines up to $12,000\lambda$ indicated a source size not more than 10 sec of arc. When similar measurements were made of distant radio sources of small angular diameter it was found that the diffraction by the ionised solar wind made the radio sources appear much bigger than their actual size. We may conclude that the true angular size of the source of Jovian decametric radiation is probably not more than about a second of arc (a diameter of less than 1/30 part of the visible disk of Jupiter).

One of the most surprising results emerging from the study of Jupiter has been the discovery of the influence of the planet Io. This is the closest of Jupiter's satellites, with an orbital period of 7 days. A new analysis of Jupiter burst data in 1964 showed that the probability of occurrence of a

decametric radio emission is greatly enhanced when Io is in two positions in its orbital path. Io obviously stimulates the emission of bursts in these regions, and it may be that Io produces some disturbance as it passes through Jupiter's "Van Allen" belts.

We shall see how the evidence for the existence of these belts has been derived from further remarkable properties of Jupiter's radio emission at short wavelengths.

Decimetric Radio Emission from Jupiter

Microwave emission from Jupiter was first detected in 1958 at a wavelength of 3.15 cm by Mayer and his colleagues at the US Naval Research Laboratory, Washington. The effective disk temperature they measured, 145 K (with an accuracy of about 25 K), was in reasonable agreement with the infrared temperature of 130 K. It was only when the radiation was measured at wavelengths of 10 to 100 cm that it was realised that Jupiter is an exceptionally strong radiator in this wavelength band. These wavelengths, of the order of tens of centimetres, are known as "decimetric" (1 decimetre = 0.1 metre). The decimetric radio emission occurs in the form of continuous radio noise (not bursts like the long-wave radiation). The effective disk temperature at 10 cm wavelength is about 600 K, and at 70 cm wavelength over 50,000 K! Obviously, then, it is a non-thermal radio source. The spectrum found by measurements at different wavelengths is shown in Fig. 4.7.

We can divide this spectrum into two parts; A the thermal radiation from the cold planet and its atmosphere at a temperature of 130 K, and B a non-thermal radiation far exceeding the thermal contribution at wavelengths above 10 cm.

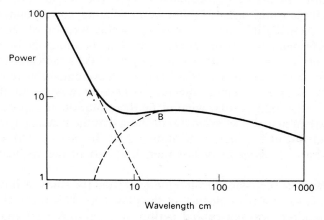

FIG. 4.7.   Jupiter's radio spectrum. (After Barber and Gower, 1965.)

One suggestion put forward to explain the strong non-thermal component was that it might be attributed to synchrotron emission from fast electrons trapped in the Jovian magnetic field in regions resembling the Van Allen belts that surround the Earth. Although the radiation from terrestrial Van Allen belts is too weak to detect, higher electron densities and higher magnetic fields may be expected on Jupiter. This idea was put to test by measurements of the size of the source made in 1960 with the interferometer at the California Institute of Technology. The source was indeed found to extend far outside the visible planet with an equatorial diameter about 3½ times that of the visible disk. The radiation was also 20 per cent linearly polarised giving further proof of the synchrotron process. Calculation of the electron energies and magnetic fields required to account for the spectral curve B in Fig. 4.7 shows that a flux of electrons with about 10 MeV energy in an average magnetic field of about 1 gauss fit well with the observations. The Jovian field will of course be considerably greater near the poles. Figure 4.8 is a pictorial representation of the "Van Allen" belts on Jupiter.

In this cross-section diagram, it will be seen that the magnetic axis is inclined at 10° to the axis of rotation. How do we know this? The answer is simple; both the intensity and the polarisation appear to "wobble" at the rotation rate of Jupiter, and this periodic variation is well explained by the 10° tilt of the magnetic axis.

Closer observations of Jupiter from space, particularly the fly-by of Voyager spacecraft in 1979, yielded much detailed information, and it is satisfying to note the confirmation of the conclusions gleaned from ground-based observations.

The Jovian atmosphere mainly consists of hydrogen and helium together with other constituents, such as ammonia, methane and water vapour. There is apparently no solid surface, and it is accepted that the bulk of the planet is in a liquid state.

The Voyager observations revealed a startling assortment of surface conditions on the Galilean satellites. Io is the scene of violent volcanic eruptions throwing out great plumes of gas and solid material to heights of more than 200 km. Although there are some mountainous regions, most of the surface of Io has a smooth covering of volcanic lava. Evidently Io's crust encloses a molten interior and suffers frequent disruption. The passage of Io through the intense Jovian magnetosphere induces powerful electric currents controlling the decametric bursts of radio emission. Of the other principal satellites, Europa has the smoothest surface appearance known in the solar system. The explanation lies in the deep layer of slushy ice that covers the satellite. Ganymede and Callisto are more solidly frozen, and Callisto's surface is a complete mosaic of craters. Ganymede with a diameter of 5200 km is the largest of all planetary satellites.

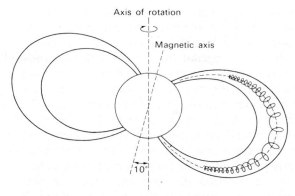

Fɪɢ. 4.8.   Illustration of Jupiter's radiation belts and spiral motion of a trapped electron.

Saturn

Saturn is the second largest planet in the solar system and has the lowest mean density 0.7 g/cm^3 (it could float on water!). Bearing a resemblance to Jupiter in size and rate of spin its cloudy atmosphere also exhibits belts and markings although less pronounced. Saturn has a comparable rotation period, 10 h 13 min near the equator and 10 h 40 min at higher latitudes. Saturn also consists mainly of hydrogen and helium together with other atmospheric constituents such as amonia and methane. There is no solid surface, and heavier elements are confined to the core.

Saturn is a weak source of long wavelength radio bursts from which, as in the case of Jupiter, the magnetic field and rotation period could be deduced. For Saturn, the field is weaker, some twenty times less than on Jupiter. There is no evidence of synchrotron radiation belts. Indeed, surrounding zones of high-energy electrons could hardly be expected to persist in the presence of Saturn's extensive system of rings.

Radiometric measurements indicated brightness temperatures ranging from 93 K at infrared to 150 K and 300 K at wavelengths of 1 cm and 1 m respectively. The higher temperatures at the longer wavelengths which penetrate lower into the atmosphere are attributable to an atmospheric "greenhouse" effect together with some internal heat generated in the planet.

Interferometric microwave studies of radio emission yielded interesting deductions on the nature of Saturn's rings. It was found that they obscured a portion of Saturn's radio emission; at the same time it was clear that the rings themselves produced negligible radiation. The results indicated that the rings are efficient scatterers of microwave radiation but not absorbers since they radiate so little. This is consistent with the interpretation of radar echo strengths leading to the conclusion that the rings mainly comprise myriads of pieces of ice typically a few cm in size.

It is noteworthy that so much basic information has been successfully gleaned from ground-based observations. Space missions have provided confirmation and revealed new data. The US Voyagers' close-up of Saturn during 1980 and 1981 exposed the magnificent detailed structure of Saturn's rings, and gave the first near view of Titan, the largest of Saturn's moons. Titan is unique as the only planetary satellite known to possess an appreciable atmosphere—found to be mainly nitrogen (together with a dense smog obscuring the surface). Voyager took 3 years to reach Saturn, and now speeds on the long journey toward Uranus and Neptune.

Saturn is the second largest planet in the solar system and like Jupiter it has a very rapid rate of spin (about 10¼ hours). Its appearance, with a dense atmosphere exhibiting belts and markings, shows a resemblance to Jupiter. Except for its greater distance from the Sun (which is of course the main source of energy for ionospheric phenomena or "Van Allen" belts), Saturn might well have been expected to bear some similarity to Jupiter in radio emission. Saturn also has, of course, a magnificent system of rings, which are believed to be made up of solid particles in orbit round Saturn and are unlikely to produce appreciable radio emission.

Uranus and Neptune

Radio emission has been detected from both these planets. Uranus is particularly intriguing as the radio brightness temperature has been rising gradually from about 160 K first recorded in 1966 to its present value of about 220 K. A possible explanation lies in the unusual disposition of the rotational axis. Unlike other planets the Uranus axis lies in the plane of its revolution round the Sun. The temperature variation is believed to be associated with the changing orientation as Uranus proceeds along its orbit round the Sun.

As we have seen, studies of the planets by their radio emission at different wavelengths can provide valuable information about the planetary surface layers and atmospheres, including ionospheres and high-energy electron belts. The data obtained in this way form one part of the planetary knowledge derived by radio astronomy. Another powerful technique for investigating the solar system is provided by radar astronomy, described in Chapter 6.

5. The Radio Sun

THE GREAT storms of radio emission from the Sun in February 1942 marked the beginning of the modern development of radio astronomy. When British Army radar stations operating at wavelengths of a few metres experienced severe jamming during late February, an investigation made by J.S. Hey (the author) led him to conclude that radio waves of amazing intensity were being radiated by the Sun, apparently due to the presence of a very large and active sunspot on the solar disk. This was the first evidence of an outburst of radio energy from an astronomical object. At this time, Forbush in the USA made another notable discovery, an intense shower of cosmic ray particles from the Sun following a large solar flare on 28th February. Later in the same year, Southworth in the USA succeeded in detecting normal solar radio emission from the quiet Sun at centimetric wavelengths. Because of the war, accounts of these important results appeared at the time in reports having only a restricted circulation. All three workers subsequently described their discoveries in scientific journals in 1946.

In this chapter, I shall describe what knowledge has been gained about the Sun by its radio emission. In many respects the radio Sun is very different from its optical appearance. At long radio wavelengths it appears not only much larger than the optical disk, but disturbances are also more intense. The radio power from the Sun can vary up to 100,000 times, as compared with a maximum change of only 1 per cent in total optical power.

Optical Features and Solar Activity

Nearly all the Sun's energy that reaches us is light and heat. The surface of the Sun called the photosphere is like an enormous furnace at 6000 deg, a higher temperature than that of an arc lamp or an oxyacetylene burner. Astronomically, the Sun is just an ordinary type of star; other stars are so much farther away that the amount of heat and light we receive from them is insignificant in comparison with the Sun. It is interesting to consider how much power is intercepted by the Earth. Measurements of the "solar constant", the flow of energy received above the Earth's atmosphere, give the value of 0.14 W/cm^2. This corresponds to 1.4kW/m^2 and the solar power falling on the sunlit atmosphere totals 1.8×10^{14} kW. Hence the Earth is continually receiving nearly 200 million million kW of power from the Sun.

As the distance of the Sun is 93 million miles (146 million km), we calculate the total power being radiated by the Sun to be 4×10^{23} kW.

We cannot examine the Sun by looking at it directly because the glare of solar light is dangerously strong. Solar telescopes are used in conjunction with filters or spectroscopes so that only part of the visible radiation is observed. The resulting image of the Sun can then be photographed. When the Sun is studied in this way it is found to possess many variable and active features. The most obvious markings are the sunspots; and although they look relatively dark because they are cooler regions of the photosphere, they are nevertheless very hot, at over 4000°C. This lower temperature compared with the rest of the photosphere is a deceptive characteristic of sunspots because they are the centres of great solar activity. Solar flares and streams of particles and X-rays originate in the vicinity of sunspots. Large sunspots can attain diameters bigger than the Earth. Solar astronomers estimate their areas in millionths of the visible hemisphere of the Sun. For example, the area of a large sunspot may be described as 5000 millionths (and hence an area of 5 thousandths of the solar hemisphere). A small sunspot may only last a day or so, but the large ones may survive for weeks. The rate of drift in the position of sunspots on the solar disk shows that the Sun is rotating with a period of 27 days. The numbers and sizes of sunspots on the Sun vary with an 11-year periodicity, known as the sunspot cycle.

An interesting characteristic of a sunspot is its strong magnetic field. In fact the commonest type of sunspot is bipolar, consisting essentially of two spots close together, one having a North magnetic pole and the other a South pole, like North and South poles of a magnet. The strengths of the fields were first measured by the astronomer Hale at the Mt. Wilson Observatory[1] in the USA. He made use of the influence of a magnetic field on atomic spectral lines, an effect discovered by a Dutch physicist Zeeman. A magnetic field alters the atomic energy levels slightly because of the effect of the field on electron rotation, and the spectral lines split into polarised components at slightly different frequencies.

The frequency separation of the displaced lines in Fig. 5.1. is the same as the gyro-frequency and is simply proportional to the field strength. It was found that sunspots have surprisingly strong magnetic fields, ranging up to about 0.4 T for the largest spots. It would take a tremendously powerful electromagnet to produce such a field over the sunspot which may have an area as big as the Earth itself! Vast electric currents must be associated with the sunspots to produce these magnetic fields.

Some fascinating phenomena occur in the solar atmosphere above the sunspot region. There are certain occasions when the hot gas suddenly brightens almost as if a huge discharge lamp has been switched on, and this

[1] The Mt. Wilson and Palomar Observatories are now known as The Hale Observatories.

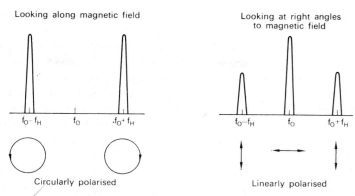

FIG. 5.1.   Zeeman splitting of spectral line of frequency f_0 in magnetic field.

event is called a solar flare. Clearly, there must have been a great surge of current in the solar atmosphere in the region above the sunspot. The discharge lamps we have for street lighting are usually either sodium vapour (yellow) or mercury vapour (bluish white) and they radiate the spectral lines of these gases. In the case of the Sun the main component of the solar atmosphere is hydrogen, and so the flare discharge produces the hydrogen line, the strongest visible line being the red hydrogen line $H\alpha$. Solar astronomers usually look for flares by observing through filters which allow through only the $H\alpha$ light. Very occasionally, flares are bright enough to stand out in white light brighter than the photosphere.

A strong solar flare often leads to disturbances in the Earth's ionosphere. Some of these occur simultaneously with the flare, so proving that they are caused by radiated waves which travel with the velocity of light. These waves are ultraviolet radiation and X-rays from the flares, and the ionisation they produce in the Earth's atmosphere at a height of about 80 km just below the base of the ionosphere often causes fade-outs of long distance radio communications. About a day or so after a large flare a more prolonged disturbance may commence in the ionosphere accompanied by changes in the Earth's magnetic field and often by displays of auroral lights. This is known as a magnetic storm, and since it occurs after the flare it must be due to something travelling from the Sun with a slower speed than electromagnetic waves. We infer that a stream of ionised gas consisting of electrons and protons must have been ejected from the solar flare. The delay of about a day indicates that the speed must be about 1000 km/sec.

Other notable features visible in the spectral line picture of the Sun are the prominences. These huge blades of solar gas appear as bright columns or arches when they are seen at the edge (often called the limb) of the Sun. On the solar disk they look relatively dark against the bright background of

the photosphere. Sometimes the prominences suddenly erupt and disappear.

In the region of the solar atmosphere around sunspots, extensive bright areas may be observed in the red hydrogen line, $H\alpha$, and these are known as plages (French for beaches). Photographs are often taken in the violet line of ionised calcium because of the high photographic sensitivity to blue and violet light. The plages are regions of hot gas in the solar atmosphere in the vicinity of sunspots, and they are actually much longer lasting than the sunspots themselves. Various types of phenomena that can be observed on the Sun are shown in the photograph (Fig. 5.2.).

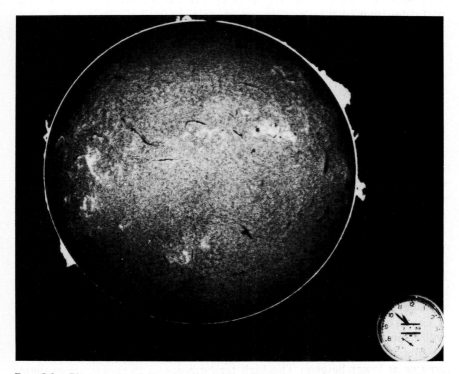

FIG. 5.2. Photograph of Sun in $H\alpha$ light showing sunspots, prominences and the chromosphere.

The Quiet Sun

The parts of the solar surface and atmosphere well away from any solar activity are known as quiet regions. The whole Sun approaches most nearly to this condition at the minimum of the 11-year sunspot cycle. There are some gradual changes in the solar atmosphere throughout the solar cycle,

but the basic characteristics remain. We will now consider the properties of the quiet solar atmosphere.

As we should expect, the atmosphere becomes more rarefied with increasing height above the Sun's surface. On the other hand the temperature of the solar atmosphere behaves in a very surprising way, as the temperature is found to rise with increasing height from the solar surface until it reaches about a million degrees. The high temperatures can be deduced from the optical spectral lines. The hotter a gas the faster the movements of the atoms, and when they collide with each other, some of the energy goes to excite the atom into high energy states. When the atom slips back into its normal state it gives out a quantum of radiation at a particular frequency, so emitting spectral line radiation. The hotter the gas the higher the energy states that occur, and the spectral line emissions indicate the high energy and enable the temperature to be deduced.

The solar atmosphere is most easily observed during total eclipses, when the Moon cuts out the glare of light from the Sun's disk and so makes it possible to distinguish the light from the glowing solar atmosphere. Close to the Sun the atmosphere appears as a fiery red ring called the chromosphere, the red colour being due to the characteristic spectral line of excited hydrogen. The chromosphere extends to a height of about 15,000 km (a great distance, but only 1/100 of the solar diameter). In the chromosphere the temperature rises steeply until the solar corona is reached. The solar corona is the outer region of solar atmosphere extending to great distances. During an eclipse the corona is seen as a diffuse white light. As we have explained, the spectral lines indicate the gas is amazingly hot, about a million degrees. A photograph of the coronal light taken during a solar eclipse in 1954 is shown in Fig. 5.3.

The optical brightness of the chromosphere and corona and the intensities of the spectral lines also enable astronomers to estimate the variation of density with height. As we shall now explain, radio observations help considerably in deducing the conditions prevailing in the solar atmosphere.

Let us consider what the quiet Sun looks like at radio wavelengths. This will depend on the electron density and the temperature at different heights. It is obvious that in any atmosphere the pressure and density are greater at lower heights. The high temperature in the solar atmosphere makes it almost completely ionised. It follows that the electron density, like the total density, increases at lower heights. We know there is a critical wavelength for the passage of radio waves depending on the electron density (see Chapter 2, page 15). In consequence, longer wavelengths cannot pass through the lower part of the solar atmosphere where the electron density is relatively great. It turns out that the lower chromosphere only allows the passage of wavelengths shorter than a few millimetres. In the upper chromosphere, only wavelengths of 10 cm or less can be propagated. Metre wavelengths are only able to travel in the corona.

FIG. 5.3. Photograph of the solar corona taken during an eclipse.

Most of the radiation at any wavelength comes from the region where the atmosphere is nearly opaque to that wavelength. Thus at millimetre wavelengths the radio emission originates in the lower chromosphere, where the temperature may be about 10,000 K (not much hotter than the photosphere at 6000 K). At 10 cm wavelength the radiation is predominantly from the upper chromosphere where the temperature is about 75,000 K. At metre wavelengths the radiation comes from the corona, where the temperature is a million degrees. Consequently, the Sun looks hotter and bigger at the longer wavelengths.

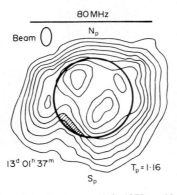

FIG. 5.4. Radio map of the quiet solar corona, July 1972, at 80 MHz. The circle marks the visible disk. The radio contours are 0.1, 0.2–0.9 T_p in units of 10^6 K. (After Dulk and Sheridan, 1974.)

1973 August 21

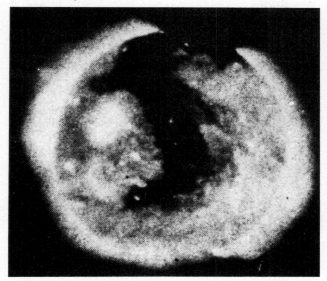

160 MHz

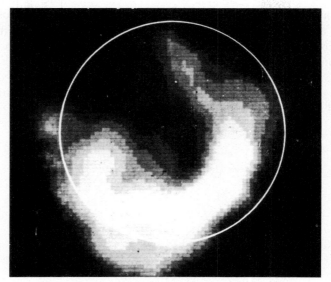

X - ray

FIG. 5.5   Radio and X-ray images of the Sun showing a coronal hole. Radio observations obtained by the Culgoora radio heliograph and X-rays by the American Science and Engineering X-ray telescope on Skylab. (After Sheridan, 1980.)

A typical radio map of the quiet solar corona obtained at 80 MHz (λ = 3.7 m) with the Culgoora radio heliograph is shown in Fig. 5.4.

Our detailed knowledge of the solar corona which was at one time practically limited to photographs taken during total eclipses is now greatly enhanced by new techniques. Not only are there available radio maps such as those obtained by the Culgoora radio heliograph, but observations from space vehicles such as the Orbiting Solar Observatory (OSO) and Skylab avoid the absorption and scattering of the Earth's atmosphere. In this way fine solar images have been obtained in white light, ultraviolet and X-rays. Figure 5.5 shows an example of the comparison between photographic representations of Culgoora radio and Skylab X-ray images of the Sun.

It has become increasingly evident that there is never a truly quiet homogeneous solar corona. Instead, zones of low intensity known as coronal holes are interspersed between brighter regions. It is apparent that the structure of the ionised solar atmosphere is dominated by the influence of solar magnetic fields. Electrons and ions spiralling about the field lines become trapped in bipolar helmet-shaped field patterns where higher densities lead to brighter emission. Between such regions are the escape zones where field lines extend outward and guide charged particles away from the Sun. These zones are the coronal holes where a fast solar wind of ionised gas escapes into interplanetary space. The solar wind is detectable in various ways, for instance by space probes or by disturbances in the geomagnetic field, and is responsible for the formation of the magnetosphere surrounding the Earth.

Radio Plages

Several radio observatories throughout the world record radio emission from the Sun regularly every day. Radio observations provide an excellent way of studying the Sun and the various types of solar activity.

When the radio power from the Sun is measured at a wavelength of say 10 cm it is found to alter gradually from day to day. This has become known as the slowly varying component (often called the S component) of solar radio emission. If we draw a graph of the daily radio intensity we find that it varies in almost exactly the same way as the total area of sunspots on the solar disk as illustrated in Fig. 5.6.

This proves that the varying component must be associated with sunspot regions. However, the lifetime of the radio emission is longer than that of the visible sunspots. Also, when the radio sources are studied with narrow radio beams (provided by grating interferometers) the emission regions are found to be larger in size and higher in the solar atmosphere than the visible sunspots. As the radio regions correspond more closely with bright plages described on page 87, they are often called "radio plages". In Fig. 5.7 a map of the radio plages is compared with that of the sunspots.

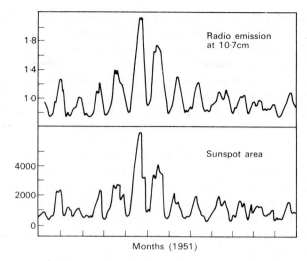

FIG. 5.6.   Comparison of radio power flux (relative units) at λ = 10.7 cm, and total sunspot area (in millionths of the solar disk) during 1951. (After Covington and Medd, 1954.)

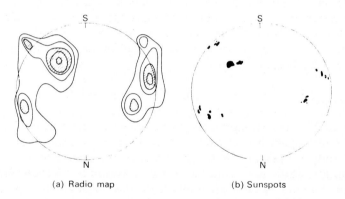

(a) Radio map   (b) Sunspots

FIG. 5.7.   Contours of radio emission at λ = 21 cm, compared with visible sunspots, 27 June 1957. (After Christiansen and Mathewson, 1958.)

The brightness temperature of the radio plages is generally about a million degrees or more. A likely explanation of the radio plages is that they correspond to high-density regions in the corona overlying the sunspots. The Swiss astronomer Waldmeier had discovered optically what he called coronal "condensations" over sunspots. We know that the temperature in the corona is about a million degrees, so if the electron density is increased in the coronal region over sunspots, making them

nearly opaque at centimetric wavelengths, then such regions will radiate at a million degrees. We may therefore explain the radio plage as the thermal radio emission from relatively dense clouds in the corona at a million degrees.

The main problem in fully accepting this hypothesis is to account for the turnover in the spectrum of the slowly varying component at very short wavelengths as shown in Fig. 5.10. It seems likely that although the radio emission is predominantly thermal, another influencing factor is also present.

Studies with higher angular resolution reveal two components, a bright core closely related to the sunspot surrounded by a diffuse halo apparently associated with the plage. The soft X-ray flux correlates well with the radio emission. X-ray photographs show bright cores surrounded by more extended structures. The radiation can be explained as thermal emission from regions of higher density trapped in the arches of the bipolar magnetic fields above sunspots.

The slowly varying component is best observed at centimetric wavelengths, from about 3 cm to 30 cm. At metre wavelengths the slowly varying component becomes submerged in the general quiet Sun radiation coming from the corona and corresponding to a million degrees. At the metre wavelengths, however, sunspots often give rise to a far more intense type of radiation called the noise storm.

Solar Radio Noise Storms

It was the occurrence of a radio noise storm at metre wavelengths associated with a large sunspot that led to the discovery, in February 1942, that the Sun could emit intense and unusual radio emission. The most striking property of the radiation is that it is so strong. Although a noise storm source is typically a fairly small area near the sunspot, covering perhaps 1/50 of the solar hemisphere, the power received at the Earth can be up to 1000 times the million-degree radiation from the whole solar corona.

Noise storms are associated with sunspot regions, but the precise conditions that make a spot noise-active are still uncertain. Large spots are the most likely to produce noise storms. The onset of a storm is sometimes apparently triggered by the occurrence of a flare. An individual noise storm may last anything from a few hours to a few days, and consists of sharp spikes (known as Type I bursts) superimposed on a general increase in radiated power.

The received intensity attains its greatest strength when the sunspot is near the centre of the solar disk, thus showing that the radiation is directed out in a beam. The waves are strongly circularly polarised due to the influence of the magnetic field of the sunspot. Estimated heights vary

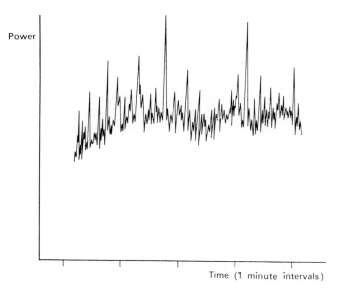

Power

Time (1 minute intervals)

FIG. 5.8.   Radio noise storm in progress with Type I bursts.

considerably—a typical height is about half the solar radius. Large bipolar sunspots are most likely to produce a noise storm. The radiation often appears to originate from a single source dominated by the leading and stronger sunspot. Some noise storms, however, show dual sources of opposite polarisations for the two main components of the bipolar spot. An example recorded by the Culgoora radioheliograph is shown in Fig. 5.9(a). A suggested model that elegantly interprets the source position in relation to the loop structure of the bipolar magnetic field is given in Fig. 5.9(b). The size of a noise storm source is typically a few minutes of arc at metre wavelengths, and the intense radiation corresponds to brightness temperatures that can exceed 10^9 K. For the Type I bursts, their short duration indicates much smaller source sizes and brightness temperatures over 10^{12} K. Evidently a non-thermal emission process is involved. Noise storms are essentially metre wave phenomena observable over a wide band as shown by the spectrum in Fig. 5.10. Despite some uncertainties in the theoretical explanation, it is generally assumed that electron streams incite strong plasma oscillations which are converted into radio waves at heights where the electron density allows escape of radio waves at the local plasma frequency.

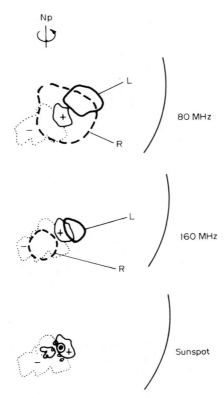

Fig. 5.9.(a)   Culgoora radio heliograph observations of left (L) and right-handed (R) circularly polarised regions of radio emission from a Type I storm on 30 October 1972. The bipolar sunspot positions are indicated below. Zones of opposite magnetic polarity are marked + and −.

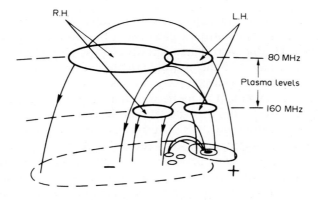

Fig. 5.9.(b)   Suggested model to account for the observations of Fig. 5.9 (a). (After Kai and Sheridan, 1974.)

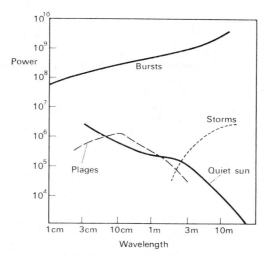

F IG. 5.10. Radio spectrum of the quiet Sun, large bursts, storms, and plages. Power flux density in units of Jy. (After Wild, 1963.)

The Solar Radio Spectrum

The most powerful types of solar radio emission are the bursts associated with solar flares. The intensities and spectra of the different kinds of solar radio emission are illustrated in the above diagram (Fig. 5.10).

We see that the bursts are by far the strongest type of radio emission from the Sun, and we will now consider what we can deduce from these exceptional phenomena.

Radio Bursts from Solar Flares

Solar flares are powerful and explosive forms of solar activity. Large flares are often accompanied at metre wavelengths by the intense radio outbursts designated Type II. But sharp bursts known as Type III normally precede outbursts, and coincide with the optical flash phase that marks the onset of most flares. Type III bursts accompany various forms of solar activity. As they are the most common of all metre wavelength bursts we shall proceed to discuss the characteristics of Type III and their interpretation.

Wild and his colleagues at CSIRO, Australia, merit the highest credit for the elucidation of the properties of solar radio bursts. The rapidity of change of the transient solar phenomena brought challenging technical problems. With interferometers and radio receivers sweeping quickly in frequency it proved possible to trace the position and spectrum of quickly moving sources. Later the Culgoora radio heliograph provided a more comprehensive view of many solar events.

Type III bursts occur as a series of sharp spikes each of a few seconds duration at the commencement of most flares. Sweep-frequency interferometry at CSIRO established that Type III radio sources move rapidly away from the Sun at speeds between a fifth and a half the velocity of light. The fast outward motion is accompanied by a drift from high frequencies to low at around 20 MHz per second. The movement in position and frequency is illustrated in Fig. 5.11.

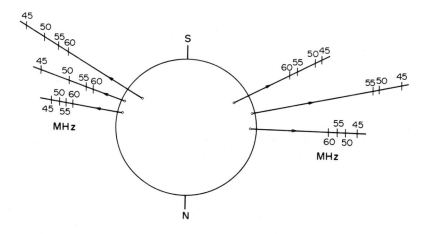

FIG. 5.11.   The drift of Type III bursts in position and frequency. (After Wild, Sheridan and Neylan, 1959.)

Since the electron density in the solar corona diminishes with increasing height, so the plasma frequency also diminishes with height. Consequently it seemed natural to interpret the bursts in terms of a disturbance moving rapidly outward from the flare, and exciting radio waves at the plasma frequency at successively higher levels in the solar atmosphere. It was inferred that the disturbance must originate from extremely fast puffs of electrons ejected at the onset of the flare. The random polarisation, the high effective temperature ($T \sim 10^{12}$ K), the radiation at the plasma frequency (and its first harmonic), all fit the idea that the moving disturbance excites plasma oscillations which are converted into radio waves. The plasma hypothesis has also been confirmed by measurements from the Helios and Radio Explorer spacecraft. The electric field of the plasma oscillations has been detected (at low frequencies of ~ 50 kHz) by Helios and Radio Explorer spacecraft as the disturbance travels out into interplanetary space. The path of the plasma source has been tracked, reaching the vicinity of the Earth some 30 minutes after it left the Sun.

Simultaneous with the ejection of the electrons generating the Type III burst, there is usually also an impulsive microwave burst (as shown in Fig. 5.16) and a burst of hard X-rays (of energy 10 to 100 keV).

Let us now picture the conditions that give rise to the solar flare and see how we can interpret the bursts of radiation that accompany the flare. It is generally agreed that flares must be initiated by an instability of the intense magnetic field associated with sunspots, and the ensuing surge of electric current. Typical helmet-shaped structures of the magnetic field above sunspot regions are illustrated in Fig. 5.12. In the outer parts the field lines are drawn outward by the solar wind. A current sheet extends between oppositely directed field lines. Any marked instability will trigger a redistribution of the field lines and tearing of the current sheet. The induced electric fields accelerate electrons to high speeds. Some escape outward along the field lines to produce Type III bursts. Others stream inward to generate the impulsive microwave bursts; for these, partial circular polarisation indicates strong fields of up to 0.1 T. As the flare develops, the disruption of magnetic fields and the associated currents cause the chromospheric brightening that constitutes the optical solar flare.

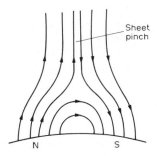

Fig. 5.12.   Typical magnetic field structure above bipolar sunspots.

Large flares exhibit a further range of dramatic phenomena attributable to the shock wave expanding from the flare explosion. Powerful radio outbursts at metre wavelengths known as Type II, and lasting between 5 and 30 minutes, accompany many large flares. By observing these outbursts CSIRO first proved that their onset occurred slightly later at lower frequencies while the source positions moved away from the Sun. As in the case of Type III bursts, the high intensity, the random polarisation, the outward movement, the drift to lower frequencies, all indicated a disturbance expelled by the flare and exciting plasma oscillations in its travel through the solar corona. A first harmonic at twice the frequency is often present. Although Type II and Type III bursts both originate from plasma

oscillations, there the similarity ends. Interferometric measurements at CSIRO demonstrated that the speed of Type II events is much slower, typically about 1000 km/s. Radio pictures obtained with the Culgoora radioheliograph, like the remarkable photograph shown in Fig. 5.13, revealed that the Type II bursts are produced by a widespread disturbance expanding outward from the flare. The speed of $\sim$ 1000 km/sec corresponds to that expected for a hydromagnetic shock wave. It is of interest to note that shock waves spreading out from large flares have also been detected optically. Another remarkable attribute of the shock wave in its passage through the solar atmosphere is its ability to trigger other phenomena in its wake such as the eruption of prominences and even other flares distant from the original flare centre.

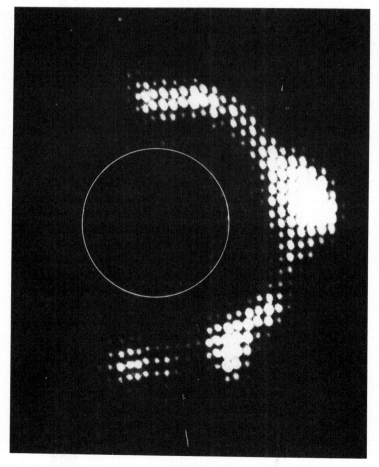

FIG. 5.13.   Culgoora radio heliograph 80 MHz picture of a Type II outburst on 30 March 1969. (After Smerd, 1970.)

It has long been known that terrestrial magnetic storms, ionospheric disturbances, and aurorae in the Earth's upper atmosphere may follow a day or two after the occurrence of large solar flares. Again the delay corresponds to a speed of travel from the Sun of about 1000 km/s, clearly coinciding with the arrival of the shock wave in the vicinity of the Earth. In addition, space satellites have recorded the shock fronts during their transit through interplanetary space.

The simplest presentation of solar bursts is on a pen recording chart, and Fig. 5.15 illustrates bursts at metre wavelengths. As well as Types II and III, sometimes a long burst covering a wide band of wavelengths occurs after the maximum of a large flare. Such long duration continuum bursts

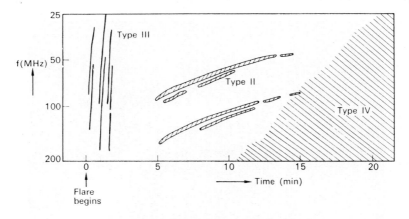

FIG. 5.14.   Dynamic spectra of radio bursts.

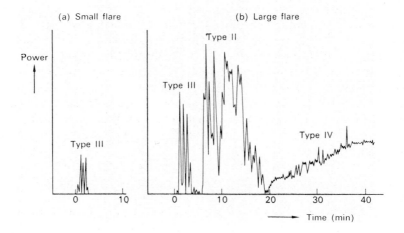

FIG. 5.15.   Pen recordings of metre wave bursts.

are known as Type IV. The corresponding spectra for these different metre wave bursts are illustrated in Fig. 5.14.

Let us now turn to centimetric wavelengths where solar radio bursts are rather different in character from metre wave bursts and are classified separately. Three main types of microwave bursts may occur in association with solar flares:

(a) an impulse burst occurring at the brightest phase of most flares;

(b) a weak gradual burst of longer duration;

(c) a large microwave outburst following the maximum phase of some large flares.

The three types are illustrated in Fig. 5.16.

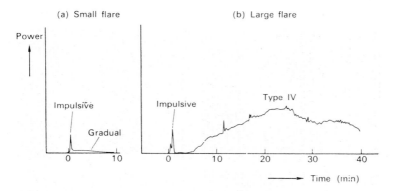

FIG. 5.16.   Pen recordings of centimetric bursts.

The microwave outbursts, often called microwave Type IV, sometimes accompany large flares and are particularly interesting because of their association with solar cosmic rays. The radio emission covers a very wide frequency band. The source shows little evidence of fast movement, but we know it is the place where high energy particles are being generated because the microwave outburst is so often followed by a stream of protons reaching the Earth with energies usually associated with cosmic rays. The particles travel with speeds not much less than that of light. For very intense outbursts, the solar cosmic rays may reach the Earth within 10 min of the start of the flare.

There seems little doubt that the Type IV radio outburst is due to synchrotron emission, and that the high energy particles are produced in the presence of solar magnetic fields. The likely explanation is that charged particles, protons and electrons, are trapped in complex magnetic fields and are then accelerated by interactions with the field. As the magnetic field declines or changes, the heavier particles, the protons, escape and are

later observed at the Earth as solar cosmic rays. Meanwhile the trapped electrons are emitting radiation by the synchrotron process.

Microwave Type IV originates from an essentially stationary and confined source at chromospheric level close to the site of the flare. At decimetre and metre waves an almost stationary continuum Type IV emission may be radiated from a large source (size ~10') in the lower corona. In this case also we assume that electrons and ions trapped in the magnetic field are accelerated to give rise to synchrotron radiation but here the region is more diffuse and lies in the lower corona.

The Culgoora radio heliograph added a new dimension to the study of Type IV radiation at metre wavelengths by revealing a spectacular variety of moving sources. It appears that these are plasma clouds ejected at speeds of a few hundred km/s containing magnetic fields and trapped

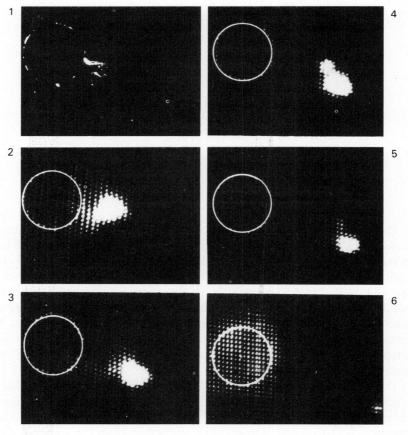

Fɪɢ. 5.17.   Sequence of moving Type IV source positions recorded by the Culgoora radio heliograph at 80 MHz on 1–2 March 1969, following the eruption of a flare prominence (top picture) photographed by the University of Hawaii Observatory. (After Wild, 1974.)

energetic electrons and ions. It is probable that in many instances the outward movement has been triggered by the passage of the flare shock wave on a region of instability. One form of Type IV can be associated with the eruption of a magnetic arch expanding at a rate of about 300 km/s, which corresponds well with the expected velocity of hydromagnetic waves (Alfvén waves) in the solar atmosphere. Another and faster form of Type IV, travelling at ~1000 km/sec, appears to develop over a wide area behind an advancing shock front, evidently as a result of compressed fields and accelerated electrons produced by the shock.

A fascinating example of another kind of moving Type IV is illustrated by the radio pictures recorded by the Culgoora radio heliograph. In this event a solar flare just out of view behind the Sun initiated a Type II radio burst and the eruption of a prominence known as a flare spray observed in Hα light by Hawaii University Observatory shown in the top photograph of Fig. 5.17. The flare disturbance was followed by the subsequent ejection of

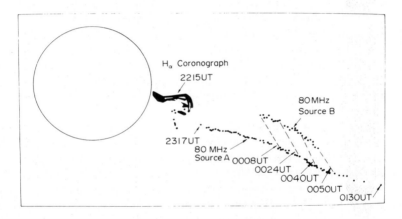

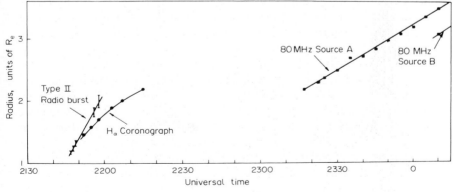

FIG. 5.18.   Trajectories and height–time graphs of the flare events shown in Fig. 5.17. (After Riddle, 1970.)

plasma cloud radiating a continuum metre wave Type IV burst. The later pictures of Fig. 5.17 show the outward progress of the source, observed on the Culgoora radio heliograph. A graph of the rates of motion of these remarkable manifestations of the flare phenomena is given in Fig. 5.18. The radiating plasma cloud detected out to a height of five solar radii splits in two parts in the later stages of its track. The complexity of solar phenomena is notorious. Nevertheless, the Sun provides a nearby cosmic laboratory offering a close scrutiny over the whole electromagnetic spectrum of an amazing range of events and a great potential contribution to the understanding of astrophysical processes.

The Outer Solar Corona and the Solar Wind

The story of solar radio astronomy would not be complete without discussing the influence of the outer solar corona and interplanetary medium on radio waves passing through them.

We know from eclipse photographs that the solar corona extends a very long way into interplanetary space. In addition, we are now aware from space-probe measurements that there is a continual outflow of ionised gas from the Sun, generally called the solar wind, extending well beyond the Earth's orbit.

The possibility of studying the outer solar corona by observing the refraction (bending) of radio waves from distant radio sources whose directions lie close to the Sun was first suggested in 1951 by Machin and Smith at Cambridge, and by Vitkevitch in Russia. They realised that the Crab Nebula would be a convenient radio source since in June each year the line of sight to the Crab passes within a distance of $5R_\theta$ from the Sun, where R_θ is the solar radius. Observations are best made with interferometers having narrow lobe spacings so that the Sun's diameter is too wide to appear on the interferometer record. It is also necessary to observe only when there is low solar activity to avoid sunspot and flare radiation. When the Crab Nebula was observed in this way, the principal effect of the solar corona was found to be not a simple bending of the ray paths but a general scattering of the radio waves by irregularities in coronal electron density. This causes the apparent diameter of the Crab Nebula to increase in size, much in the same way as when we look at a light through a frosted or rough glass surface. The Crab Nebula actually subtends an angle of about 5', but the apparent size at 8 m wavelength at a distance of $10R_\theta$ from the Sun was found to be 15' rising to 45' at $5R_\theta$. The apparent increase in source size reduces the amplitude of the recorded interference fringes. The position of the Crab Nebula relative to the Sun on different dates in June, and the effect of coronal scattering on fringe amplitude is illustrated in Fig. 5.19.

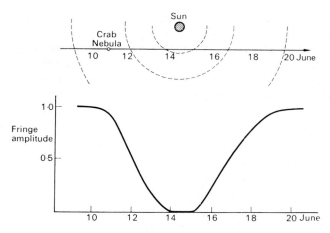

FIG. 5.19. Decrease of interference fringes from Crab Nebula due to solar coronal scattering. From observations during June 1956 at $\lambda = 3.7$ m with an interferometer spacing of 105λ. (After Hewish, 1958).

The scattering is caused by variations in electron density in the outer corona. The corresponding changes in radio refractive index are proportional to λ^2, where λ is the wavelength. The scattering is therefore much greater at longer wavelengths. In fact, if the scattering is very large, there can be a fall of received signal as well as a reduced visibility of the interference fringes. The scattering process is illustrated in Fig. 5.20.

Experiments made with interferometer baselines in various directions have led to the interesting deduction that the ionisation irregularities are elongated and aligned almost radially out from the Sun. The alignment of

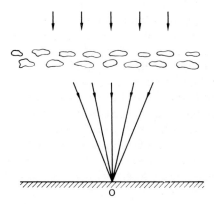

FIG. 5.20. Radio waves scattered by irregularities reach the observer O in different directions.

the irregularities results from the action of the solar magnetic field. As the electrons spiral around the lines of force they are prevented from diffusing out sideways. The irregularities therefore travel and diffuse mainly along the lines of force. The radio picture of the alignment of ionisation in the vicinity of the Sun shown in Fig. 5.21 ties up well with the appearance of visible plumes and streamers in the corona.

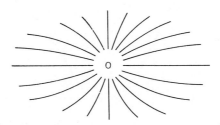

FIG. 5.21.   Illustration of the field pattern in the outer solar corona.

The radio scattering is strongest in the equatorial plane, indicating that the form of the electron density distribution is elliptical. The magnitude of the scattering varies with the sunspot cycle as we may expect. The scattering falls off with increasing distance from the Sun, and the results fit well with space probe results indicating about 5 electrons/cm^3 at a distance of 1 A.U.

The scattering properties of the outer solar corona and solar wind are also manifested in another way. In 1962 it was found in observations at metre wavelengths of sources of very small diameter (less than seconds of arc) the scattering produces scintillations of amplitude. This is analogous to the twinkling of stars caused by irregularities of refractive index in the Earth's atmosphere. The difference is that the optical fluctuations are due to variations in atmospheric temperature and water vapour, whilst the radio scintillations are caused by variations of interplanetary electron density. In both cases the speed of winds moving the irregularities across the field of view determines the rate of fluctuation that is observed.

The reason for the changes of amplitude is an effect of interference between the scattered wavelets. The irregular changes of refraction bend the waves so that they may reach the observer from a variety of directions. The result is that waves scattered in different directions combine, sometimes adding in phase to give a large signal, and sometimes out of phase to cancel each other. The observed signal is altered accordingly.

If the source is larger than a certain size the fluctuations are smoothed out, as illustrated in Fig. 5.22.

Suppose the average size of an irregularity is L and its distance from the observer is D. Then if the source subtends an angle θ equal to or greater

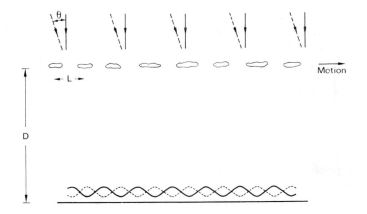

FIG. 5.22. Fluctuations smoothed out by combining radiation from different parts of the source.

than L/D the amplitude fluctuations are smoothed out because the maxima from one part of the source cover the minima from another.

Observations of fluctuations have proved a valuable means both of distinguishing sources of very small size, and of finding the scale and structure of the interplanetary ionisation due to the solar wind. Recordings at widely separated observing sites show a pattern of fluctuations received first at one site and then at another as the irregularities move with the solar wind. This is rather like sunlit cloud pattern moving across the Earth's surface. The radio observations indicate interplanetary wind speeds of 300 to 500 km/sec, and irregularities of the order of 300 km in length. The path of motion of the solar wind is similar to the trailing flow from a rotating hosepipe as shown in Fig. 5.23 as a consequence of the Sun's rotation.

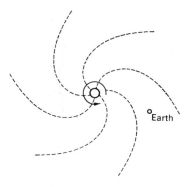

FIG. 5.23. Effect of solar rotation on the geometry of the solar wind.

The irregularities flowing along these paths are held together by magnetic fields, for they undoubtedly take some of the solar magnetic field with them. This is sometimes called the "frozen field" effect. It could equally be described as due to persistent electric currents because magnetic fields are produced by electric currents, that is, by moving charges. In low density plasma there are so few collisions between particles that currents continue to flow unimpeded.

From our investigations of the radio Sun we have become aware of a more extensive picture of solar phenomena, some of which may be invisible or only dimly seen. Radio wave observations make particularly important contributions to solar physics in two ways. We learn a great deal about the solar corona and solar wind. Radio also provides a powerful means of detecting disturbed regions and fast-moving particles.

6. Radar Astronomy

WE NOW turn to radar observations which provide a different way of studying the solar system. Radar methods were first applied to astronomy in 1945–1946 when Hey and Stewart in Britain investigated echoes from meteor trails, and when radar echoes from the Moon were obtained in the USA and Hungary.

In radar astronomy we send out radio waves, and then detect the radio echoes returned from the astronomical objects. Such methods can only be applied at comparatively short astronomical distances, for the following reasons. Firstly, radio waves have to traverse the two-way path, to the target and back again. Radiated power flux falls off as $1/R^2$, where R is the distance, and the power flux scattered back also falls off as $1/R^2$. Consequently the returned power flux diminishes as $1/R^2$ times $1/R^2$, that is $1/R^4$. With increasing distance, the returned power therefore falls off rapidly so that the echoes from very distant objects are too weak to detect. Secondly, the powers that can be generated in radar transmitters are puny compared with those produced by astronomical sources. Thirdly, only a tiny fraction of the power transmitted to an object is reflected back towards us. Lastly, even if we could achieve long ranges to the stars and galaxies, it would take many years (or even millions of years!) before the echo came back. For the above reasons, we conclude that radar astronomy is impracticable beyond the solar system.

Within the solar system, radar has a valuable role to play and possesses two very important assets. It enables us to measure distances and speeds very accurately. The range of an object is indicated by the time taken for the echo to return, for it is well known in radar that each microsecond (millionth of a second) corresponds to 150 m of range[1]. The speed the object approaches or recedes from us is given by the Doppler shift of the frequency of the reflected signal.

Before we describe the results obtained, we must mention certain factors affecting the strength of the observed echo. The first factor is the receiver bandwidth. In radar we make this as narrow as possible, just keeping it wide enough to contain all the frequencies in the radar pulse together with any Doppler shifts produced by the motion of the reflecting object. By having a narrow bandwidth, we keep to a minimum the radio noise produced by the receiver or entering from outside sources. Good radar

---

[1] This is easily worked out. Radio waves travel at 3×10^8 m/sec. In 1 μsec, the waves therefore travel 300 m, equal to a distance of 150 m there and back.

detection depends on having the echo signal large compared with the noise level.

Now consider how much power is reflected back from the object under observation. Any radar target is said to have an equivalent echoing area or radar cross-section. It is defined as the area of an isotropic scatterer normal to the illumination that would yield the observed echo intensity. It can readily be proved that the echoing area of a perfectly reflecting sphere, if it is large compared with the wavelength, is given by the cross-sectional area, πr^2, where r is the radius. If the sphere is a poor reflector, for instance, if it only reflects a tenth of the power falling on it, then its echoing area is $0.1\pi r^2$. Thus if we measure the total echo strength from a planet, we then know its reflection coefficient q and hence we have a good clue as to its composition. For example, water would give a strong reflection, with q almost 1, while dry, solid rock would have a reflection coefficient of only $q \approx 0.15$. If we use short pulses we can compare the echoes from different parts of objects and make deductions about the slopes and roughness of the surface. We will illustrate this by considering the radar investigation of the Moon.

Radar Echoes from the Moon

Let us suppose we have a powerful radar looking at the Moon. The lunar disk subtends ½°, and the radius of the Moon is 1740 km. The type of radar we would use to measure the total echoing area of the Moon would therefore have a beamwidth larger than ½° to encompass the Moon, and a pulse length longer than 11.6 msec. Observations at various wavelengths between 8 mm and 8 m indicate an echoing area of about $0.07\pi r^2$. A reflection coefficient of 0.07 is appreciably less than for solid, dry terrestrial rocks. This has suggested that the Moon's surface is either porous (like pumice stone) or of a light or granular structure.

We may study the Moon in more detail by examining the echo from different ranges on the lunar surface with short pulses. The first reflection comes back from the nearest point on the Moon as illustrated in Fig. 6.1(a).

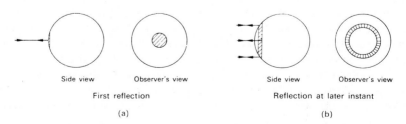

Side view   Observer's view   Side view   Observer's view

First reflection   Reflection at later instant

(a)   (b)

FIG. 6.1.   Radar reflection from the Moon.

At a later instant the echo is coming back from a region near to the limb of the Moon. The region at a given range is an annular ring as shown in Fig. 6.1(b).

The returned echo from the nearest point is very strong, and tails off rapidly with increasing distance as shown in Fig. 6.2.

If the Moon had been perfectly smooth, the echo would have come entirely from the nearest zone. If the Moon had been entirely rough we would have had strong echoes right up to the outer limb. We illustrate smooth and rough reflection in Fig. 6.3.

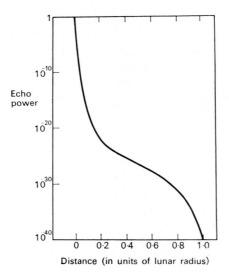

FIG. 6.2.   Dependence of echo power on distance from the nearest point of the moon. From observations at $\lambda = 68$ cm. (After Evans and Pettengill, 1963.)

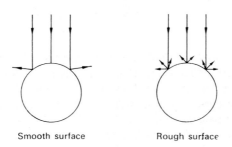

FIG. 6.3.   Reflection from smooth and rough surfaces.

The actual results shown in Fig. 6.2 tell us that the radar scattering from the Moon is neither very smooth nor very rough but intermediate between the two. From the shape of the curve we can deduce that the average slope, at the scale of the radar wavelength, is about 1 in 7. At shorter wavelengths (cm and mm) the echo in the outer zones becomes stronger, indicating small-scale roughness. We know, too, that at optical wavelengths the scattered light corresponds to a rough surface.

A further step in discriminating different parts of the Moon's surface was achieved at the Lincoln Laboratory, USA, by examining the Doppler shifts of echoes at different ranges. The method can be explained in the following way. Although the Moon keeps the same face towards the Earth it appears to have a slight rocking motion, known as libration. This means that at any instant it appears to be rotating slightly to and fro. The rotating movement of the surface produces Doppler shifts depending on the position of the Moon, and we can divide the surface into strips each corresponding to a particular Doppler shift. As there are also circular zones corresponding to different ranges, we have a means of picking out particular regions of the surface as illustrated in Fig. 6.4.

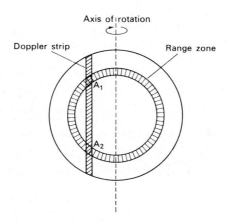

FIG. 6.4. Zones of range and Doppler shift.

A_1 and A_2 are regions on the Moon corresponding to a particular range and Doppler shift. Further refinement is achieved by interferometry. By using two receiving aerials to form an interferometer the echoes from A_1 and A_2 can be distinguished separately. Remarkable resolution has been attained by ground-based radar employing the technique combining Doppler frequency, range and interferometry.

An excellent example of a radar photograph of a region of the Moon derived at centimetric wavelength by this method is presented in Fig.

6.5(a). The signals transmitted from the Haystack radar, shown in Fig. 3.7(f), were received after reflection from the Moon at both the Haystack radio telescope and another aerial at Westford 1.2 km away so forming an interferometer system. Not only does Fig. 6.5(a) portray the same detail as that of an optical picture with a resolution 2 km but the method also determines heights to within 100 m; a map of the height profiles for the same region is shown in Fig. 6.5(b).

Radar Echoes from Venus

The planet that comes nearest to the Earth is Venus. At its closest approach its distance is about 40 million km, over a hundred times more distant than the Moon. The inverse fourth power distance law ($1/R^4$) would make it 100 million times more difficult to detect than the Moon, although the greater dimensions of Venus reduce this factor to about 10 million times. Thus a tremendous increase in sensitivity is required compared with the Moon. It is not surprising that after the first radar observations of the Moon, it was 15 years before echoes were successfully obtained from Venus. How was the required improvement in performance achieved? Three main factors were responsible: (a) larger radio telescopes giving more directive transmission and reception, (b) low noise receivers such as masers, and (c) long integration, that is, the adding together of a large number of echoes.

The time taken for a radio signal to travel to Venus and back at closest approach is 5 min (as compared with 2½ sec to the Moon). The usual method is to transmit a series of pulses for just under 5 min, and during the next 5 min receive back the echoes, and keep on repeating the procedure. If the received signals are recorded on magnetic tape, the integrating process of adding many echoes together can be carried out at a later time. Echoes from Venus were first obtained with certainty in 1961 at Lincoln Laboratory, USA.

One of the most important results emerging from this experiment was the accurate determination of the astronomical unit of distance, that is, the mean distance of the Earth from the Sun. The positions and velocities of the planets are known quite accurately relative to the Earth and the Sun. Thus a single accurate planetary radar measurement is sufficient to determine the true distance scale and hence the astronomical unit (A.U.).

Prior to the radar observations of Venus, the A.U. determined by optical methods was not known more accurately than 1 part in 1000. The first reported radar observations of Venus in 1958 and 1959 suffered from inadequate signal strength and the measurements subsequently proved to be erroneous. The diagram (Fig. 6.6) shows that by 1961 mutually consistent radar results were being obtained, leading to a far more accurate value for the A.U.

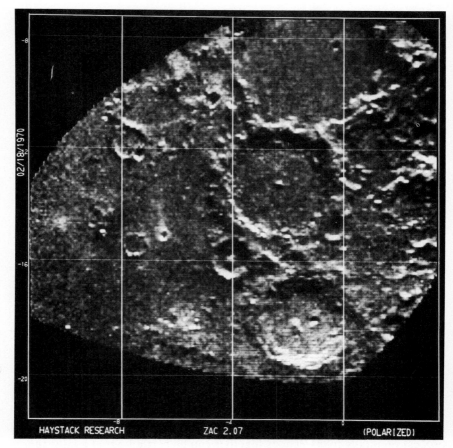

FIG. 6.5.(a)   Radar-interferometer image of a region of the Moon showing the lunar craters Alphonsus, Ptolemaeus, Arzachel and part of Mare Nubium. (After Zisk, 1972.)

On the basis of the radar measurements the International Astronomical Union decided in 1976 to adopt 149,597,870 km as the officially recognised value of the A.U. The accurate determination of the A.U. is important because it is the essential unit on which all other distances depend. For example, the distance corresponding to the parsec used for stellar distances depends on the A.U. Also, space probe missions to planets could not be achieved without an accurate knowledge of distances.

The characteristics of the radar echo from Venus tell us about the surface conditions and rotation of the planet. The surface is optically obscured by the opaque atmosphere. The radar reflectivity of the whole planet is about 16 per cent; this is a stronger reflectivity than that of the

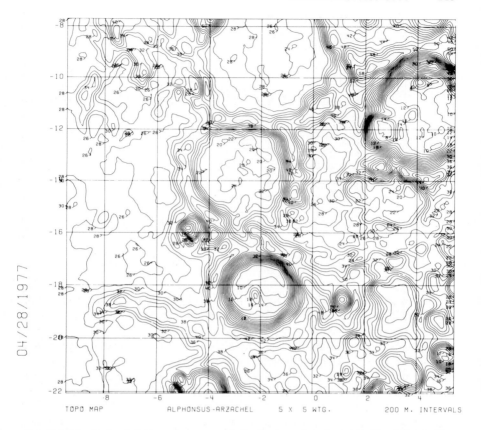

04/28/1977

FIG. 6.5.(b)   Radar height contour map of the lunar region shown in Fig. 6.5 (a). The
contour interval is 300 m. (After Zisk, 1972.)

Moon, and the value is consistent with dry rocks similar to those found on
the Earth's surface, or with dry sand. As we move towards the limb, the
radar echo falls off more steeply as compared with the Moon, which shows
that Venus has a smoother surface.

The determination of the rotation of Venus excited much interest. All
the planets, as they move in their orbits, go round the Sun in the same
sense, anti-clockwise as viewed from the North. If a planet spins about its
axis in the same sense, its direction of rotation is called direct. If it rotates
in the opposite sense it is called retrograde. It has always been thought that
all planetary spins except the distant Uranus were direct. Radar observations show that Venus has a slow retrograde rotation with a period of about
150 days. This is the sidereal rate (with reference to the stars), and when
combined with the orbital motion, it is found that the solar "day" on Venus
lasts about 120 days.

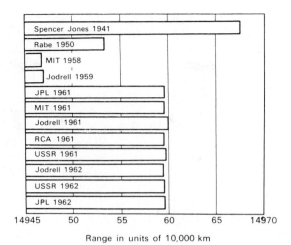

FIG. 6.6.   Measurements of the astronomical unit of distance.

Although mapping the surface of Venus is obviously difficult because of its small angular size, a valuable preliminary survey was made with a transmitter in the 300 m Arecibo radio telescope by exploiting the same techniques as those used in radar observations of the Moon. In addition a few small areas were photographed optically by the Venera landing craft, revealing a rocky surface. Radar clearly offers the best hope of obtaining a more general picture of surface topography through the dense cloud cover. To this end, the U.S. Pioneer Orbiter spacecraft was fitted with a radar to scan the surface. There are evident advantages in radar observations from close range. Only modest power is required (20 watts in the Pioneer radar), and the spacecraft orbits can be arranged to cover most parts of the surface. Despite Pioneer's highly eccentric orbit (primarily designed to permit a wide range of atmospheric studies) Pettengill and his colleagues in the USA, who were responsible for the radar equipment, successfully derived a near-global map of the surface. Altimetric measurements gave heights with an accuracy better than 200 m averaged over areas of the order of 10 km diameter. The observations revealed a great variety of features, plateaus, mountains, and evidence of large volcanoes and meteor craters in the past history of the planet. The most striking prominence is a mountain in the region named Maxwell rising to a height of 11 km, 2 km above that of Everest.

Mercury

For all the planets except Venus the sense and period of rotation have been determined by visual observation. When radar echoes were first obtained

from Mercury (in the USSR in 1962 and in the USA in 1963), a rotation period of 59 days was deduced. This was a surprising result because optical observations had been regarded as indicating that the planet always turns the same face to the Sun, therefore having an 88-day direct rotation exactly equal to the orbital period. The optical data was at once re-examined and it was then realised that in the original optical analysis, alternative interpretations of the data had been ignored. It is now agreed that the correct rotation period is 59 days as found by radar. A solar "day" on Mercury in this case corresponds to about 200 days. As all the surface is periodically subjected to solar radiation we are able to account for the radio temperature measurements (see Chapter 4) indicating a warm sub-surface on the dark side.

Mercury is the smallest of the planets and its direction close to the Sun has led to some difficulty in observation. Radar has provided a useful source of information on surface topography indicating hills and valleys with height differences of about 1 km. The reflectivity is similar to that of the Moon.

Other Radar Planetary Studies

Planetary exploration has been greatly advanced by the advent of space research. As we have seen, space-borne radar has an important role in planetary observations. In contrast, ground-based observations have achieved many outstanding successes, and have the advantages of comparatively economical and more continuous operation.

In concluding the discussion of planetary radar several further points of interest should be noted. Mars has not been a subject of very extensive radar study, firstly because the echoes received at the Earth at closest approach are 200 times weaker than those from Venus, and secondly the surface is amenable to direct optical view. Jupiter is an enormous planet but the absence of a clear-cut surface and absorption prevents appreciable radio reflection. Similar restrictions apply to Saturn. Improvements in sensitivity have, however, brought many planetary objects within the scope of detection. Radar echoes have been obtained from the four principal satellites of Jupiter. In fact, the icy surfaces were first inferred from unusual characteristics of the radar echoes. Various minor planetary objects, a few hundred km in size, left over after the formation of the main planets, have been detected. These include Icarus, Eros, Ceres, Betulia and Toro.

The radar detection of Saturn's rings afforded a striking example of unexpected results leading to highly significant inferences. Twice during the 29 years of Saturn's orbital period the rings are tipped at maximum inclination to the line of sight affording a favourable aspect for radar observation. An attempt was made to record radar echoes at the optimum

aspect (during 1972–3) using a 400 kW transmitter at 12 cm wavelength in a 64 m telescope at the Jet Propulsion Laboratory, USA. The time taken for an echo to return to Earth is about 2¼ hours. Although previous calculations had suggested that the rings might well be beyond detection, surprisingly strong echoes were in fact obtained, and later confirmed by measurements at λ = 3.5 cm. The echoing area corresponded to more than 60 per cent of the visible area of the rings. The results implied multiple reflections within and between particles of very low dielectric loss and dimensions mostly of at least a few cm. The only feasible explanation is that ice is the main constituent of the rings of Saturn.

Radar Echoes from the Sun

Radar observations of the Sun present a different problem because of its highly ionised atmosphere. It might be thought that reflection close to the visible Sun could be obtained at short wavelengths, but actually the dense ionised lower levels of the solar atmosphere would absorb all the incident radiation. The only feasible way of obtaining radio echoes is at long wavelengths because the radio waves are then reflected high in the solar corona without much absorption. Although the solar corona may have a large echoing area at long wavelengths, observation is severely hampered by the presence of strong radio noise emission from the Sun, reaching great intensity during solar activity. The first successful demonstration of radar echoes from the Sun was achieved in the USA at a wavelength of about 10 m. Since then several studies of solar radar echoes have been made, but analysis is a complex problem, involving appropriate allowance for refraction, absorption, and scattering from irregularities. The echoes at λ = 10 m were found to occur at a height of 1½ times the solar radius, with a wide Doppler spread indicating large-scale movements of the ionised atmosphere, and a mean Doppler shift as expected from an outward-flowing solar wind.

Meteors

When we see a meteor producing a bright streak of light in the sky it is hard to believe that it is only a tiny speck of matter entering the earth's atmosphere. The light results from the enormous speeds being sufficient to produce glowing ionised gas. The heat developed by impacts with air molecules normally burns up the meteor at a height of around 100 km (60 miles). Meteors occur in all sizes but most are very small particles; the large objects that reach the ground are called meteorites and are fortunately rare. It has been calculated that the average rate of occurrence of meteorites of more than 50,000 tons is 1 per hundred million years!

The speeds of meteors are between 11 km/sec and 72 km/sec. The lower limit of 11 km/sec is set by the velocity of free fall due to the pull of gravity. The upper limit is set by the velocity of the Earth in its orbit, 30 km/sec, plus the maximum meteor velocity of 42 km/sec of an approaching meteor. The latter value, 42 km/sec, is obtained by assuming that all meteors are members of the solar system. If a meteor in interplanetary space has a lower velocity than 42 km/sec it is held in an orbit round the Sun. If it has a higher velocity, it escapes from the constraint of the gravitational attraction of the Sun and moves out of the solar system.

Meteors entering the Earth's atmosphere strike air molecules, causing the meteor surface to become so hot that the meteor evaporates. The ejected atoms have the high velocity of the meteor, and electrons are knocked out by further collisions with air particles, leaving an ionised trail. The glowing ionised gas is often visible—the "shooting star" we occasionally see at night. There are more visible meteors than we imagine, and a trained observer may see about 4 to 10 per hour. This means about 10^8 visible meteors per day over the whole surface of the Earth.

The sizes of most meteors are extremely small. Even a very bright meteor of 1st optical magnitude has a diameter of only a fraction of a millimetre. The smaller the size the more meteors there are, although most are too faint to see. The total meteoric mass falling into the Earth's atmosphere has been estimated at about 9 tons per day.

Meteors can be divided into two classes; the sporadic meteors that come in all directions, and the shower meteors that appear at particular times of the year coming from certain directions in the sky called the meteor radiants. The meteor showers are named after the constellations in the direction of the radiants. It is evident that the shower meteors are streams of dust and particles spread out along confined orbital channels. It is known that many of these streams are the debris left by comets. When the Earth's orbital path intercepts such a stream, a meteor shower is observed. Figure 6.7 illustrates the Earth approaching the well-known Perseid meteor stream which it passes through in August each year.

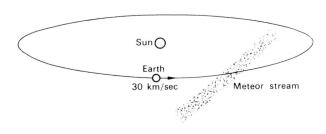

FIG. 6.7.   The Earth passes through the Perseid meteor stream each August.

The first established radar observations of meteor trails were made by Hey and Stewart in 1945–1946 when they proved that radar echoes could be obtained from meteor trails and they determined the radiants of meteor showers. Some of these were daytime showers and therefore previously unknown. Meteor velocities were then measured by Hey, Stewart and Parsons, and radar methods soon became firmly established in meteor astronomy. We will now describe briefly the properties of radar echoes from meteor trails, and some of the methods for deriving the astronomical data.

The meteoric particles are normally far too small for their echoes to be picked out on the radar screen. What we observe is the echo from the ionised column produced by the meteor, and this is strongest when we look at right angles to the trail. The echo is produced by the scattering of the radio waves from the electrons, and the reflected signal follows the usual laws of reflection. We know that a mirror reflects light back when it is perpendicular to the light path.

This directional property of reflection was utilised by Hey and Stewart in 1945 when they first determined the radiants of meteor showers. They used three radar sets at 4 m wavelength with elevated beams looking at different azimuths. The sites are illustrated in Fig. 6.8.

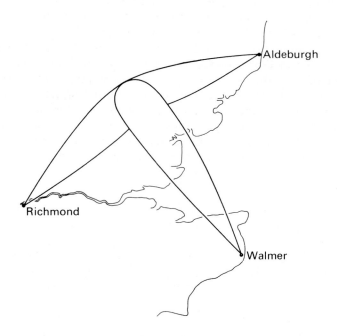

FIG. 6.8.   Determination of meteor radiants by Hey and Stewart, 1945.

They found that the rate of occurrence of echoes came to a peak at different times of day on the three radar equipments. The peaks corresponded to the times when the meteor radiants lay perpendicular to the aerial beams. Using this information they calculated the positions of the radiants. This experiment is now mainly of historic interest, but it was the first radar determination of meteor radiants and a demonstration of the aspect sensitivity of meteor echoes.

The meteor trail initially forms as a narrow, straight column of ionisation. The trail then expands and electrons are gradually lost by recombination or attachment. The echo therefore rapidly fades in strength. Since we are observing an ionised gas we know that we must use a long wavelength to obtain a strong reflection, and wavelengths of several metres are very suitable. The echoes are by far the strongest if the electron density exceeds the value for total reflection from an ionised gas. It may be shown that this requires the column to have more than 10^{14} electrons/m of trail length.

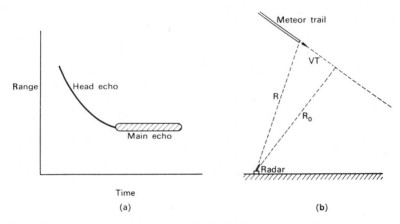

FIG. 6.9.   Determination of meteor velocities by Hey, Parsons and Stewart, 1946.

The first radar measurements of meteor velocities were made by Hey, Stewart and Parsons during the 1946 Giacobinid meteor shower when they noticed a faint approaching echo appearing before the main trail echo had formed. This was the echo from the head of the ionised column, that is, from the ionisation in the vicinity of the approaching meteor. Figure 6.9(a) shows the appearance of the recorded echo.

Figure 6.9(b) illustrates the radar observing a meteor having velocity V. Applying the Pythagoras theorem we see that the range R at time T before the meteor reaches the nearest distance R_0 is given by $R^2 = R_0^2 + V^2T^2$. The velocity V is then easily found from the range-time plot of the head

echo in Fig. 6.9(a). For the Giacobinid meteors the speed was found to be 23 km/sec.

The head echo is often too faint to detect, and a method for measuring velocities developed by Davies and Ellyett at Jodrell Bank is now more often used. Their technique depends on a characteristic modulation of the signal due to an interference effect as the meteor passes at minimum range R_0. The effect can be understood by considering the echo in two parts; one from the first half of the trail and the other from the second half just beginning to form as the meteor passes beyond minimum range. To start with, the echo from the newly-formed second part is in phase with the echo from the first half of the trail. As the meteor moves on, the waves echoed from the newly-formed part get out of step because of the increased range. Later they come in step again and so on. In consequence the total echo amplitude oscillates near the point of minimum range as shown in Fig. 6.10.

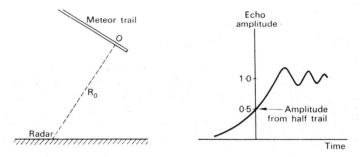

FIG. 6.10.   Amplitude variation during formation of meteor trail. (After Ellyett and Davies, 1948.)

The radar study of meteor trails is now a fine art. The radiants and velocities of individual meteors can be measured very accurately and the orbits about the Sun can be deduced. As an additional bonus, the subsequent drift of the trails after formation tells us the wind speeds and directions in the Earth's high atmosphere.

7. The Galaxy

THE DETECTION of radio emission from the Milky Way by Jansky in 1933 was the first observation in radio astronomy. Later Reber mapped the distribution in more detail showing clearly the association of the radio map with the Milky Way, and hence with the general structure of the Galaxy. Reber thought the radiation might be produced by interstellar ionised hydrogen. To solve the problem of the origin of the radio emission we must examine closely the characteristics of the general galactic radiation and its relation to the structure of the Galaxy. First let us look at the optical picture.

Optical Evidence of Galactic Structure

There is a great temptation when describing the structure of the Galaxy to refer instead to Andromeda and other relatively near spiral galaxies outside our own. The reason is that is is hard to decipher the structure of our Galaxy because we are inside it. The best view of the density and movements of traffic on congested roads is obtained by surveying the situation from a helicopter. Sorting out our Galaxy from the Earth is like trying to unravel a maze from the inside. The first idea that our Galaxy might have the form of a flat disk divided into spiral arms came from looking at Andromeda, 2 million light years away, shown in Fig. 7.1.

Since the time of the astronomer William Herschel (1738–1822) who made the first detailed study, attempts have been made to deduce the spiral structure and dynamics of our Galaxy directly from the disribution of stars and their relative motion. The stars are often described as "fixed", yet when measurements are made at intervals of years, small changes of position are found. These displacements indicate transverse components of movement. At the same time, the Doppler shifts of spectral lines tell us the speeds with which the stars are moving towards or away from us. A great deal of important information about the Galaxy has been derived in this way. It has been shown that the galactic disk has a diameter of about 150 thousand light years. The Sun is located about 30 thousand light years from the centre. Although the stars have random movements of the order of 10 km/sec with respect to each other, there are also systematic movements at much higher speeds. For example, the Sun together with the near stars, the "local group" as it is called, are all moving at about 200 km/sec, apparently

Fig. 7.1.   The Andromeda Nebula.

towards the direction of the constellation of Cygnus. We can interpret this wholesale movement as due to the general rotation of the galactic disk.

The principal optical components of the Galaxy are illustrated in Fig. 7.2. The galactic disk contains most of the stars, including the large hot new stars of types O and B and the glowing ionised hydrogen surrounding them. There is also another population of stars concentrated towards the central part or nucleus of the Galaxy, producing the central bulge shown in Fig. 7.2. In addition there is a widely dispersed distribution of older types of stars with little or no concentration to the galactic plane. These are contained roughly within a large spherical volume called the galactic halo.

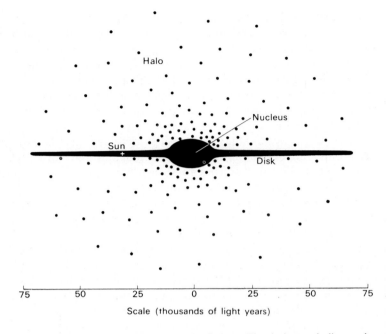

FIG. 7.2.   Illustration of the components of the Galaxy. The dark area indicates the region
where most of the stars, gas and dust are concentrated.

Although a great deal of knowledge about the components of our
Galaxy has been derived optically, it was not until 1951 that a picture of
spiral structure had emerged from the optical data. Obscuration by dust is
the greatest obstacle to visual observation. Radio astronomy has now
provided a much clearer picture of the form of the Galaxy. Radio waves
are not hindered by the dust, and the discovery of the radio spectral line of
atomic hydrogen at 21 cm wavelength opened a new era in the study of
galactic structure. The spectral line has particular importance because
motion of the source produces a Doppler shift enabling the velocity to be
deduced. Let us now consider what information about the Galaxy has been
obtained by radio methods.

The 21 cm Hydrogen Line

The element that is most prevalent throughout the universe is hydrogen.
The Galaxy contains a vast accumulation of gas, a considerable part of
which has condensed under the force of gravity to form the stars. In 1945
the Dutch astronomer van de Hulst predicted that a radio line at 21 cm
wavelength corresponding to a hyperfine transition of atomic hydrogen

should be detectable in the Galaxy. The nature of this transition is explained in Chapter 2. The interstellar gas is so rarefied that in the laboratory we should describe it as almost a perfect vacuum; on the average there is only about 1 atom per cubic centimetre. The atomic transition producing the line is also a remarkably infrequent occurrence. In fact, the average lifetime of an atom in the higher hyperfine state is 11 million years before it emits a quantum of radiation. Nevertheless, the extent of the Galaxy is so vast that sufficient hydrogen is present for the spectral line to be observable. It was first detected in 1951 by three research groups independently in the USA, Holland and Australia. The group at Leiden, in Holland, started a systematic survey of the distribution and motion of the galactic hydrogen in the northern part of the sky. Subsequently, a similar survey made in Sydney, Australia, filled in the observation of the southern part of the sky. A remarkable picture of the spiral structure of the Galaxy has emerged.

Let us examine briefly how a spiral arm may be observed by the 21 cm line radiation. In Fig. 7.3, S marks the position of the Earth and solar system, and the shaded curve represents part of a spiral arm. Looking in direction B, the gas of the spiral arm is approaching directly towards S, so strong Doppler-shifted emission is recorded. In direction A both the path in the spiral arm and the components of velocity toward S are less. Direction C is outside the spiral arm so the signal has fallen to zero. Observations in the different directions therefore give a clear indication of extent and velocity of the spiral arm. If the arm rotates in the opposite direction the shifts are reversed. Allowance must of course be made for the motion of the Sun when interpreting the Doppler shifts.

In addition to the 21 cm emission, valuable supplementary information can often be obtained by observing the line in absorption when looking in the direction of a strong discrete source. We know for example that if on a sunny day a thin cloud covers the Sun, we receive less light and heat. In the same way when a radio telescope is directed to a strong radio source, the intervening hydrogen can be detected by the absorption it produces at 21

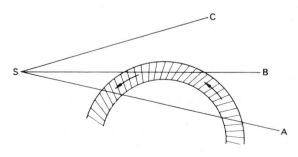

Fig. 7.3.   Observation of a spiral arm.

cm wavelength. This is a sensitive way of detecting interstellar hydrogen, because the absorption shows up quite easily when looking towards strong sources. In addition, when looking at sources within the Galaxy, we know that any absorbing hydrogen must be closer to us, and this is a useful aid in finding roughly where the hydrogen lies in the Galaxy.

Let us see some of the results that have emerged from the 21 cm line measurements. The spiral arm structure shown in Fig. 7.4 has been obtained by combining results obtained in Holland and Australia.

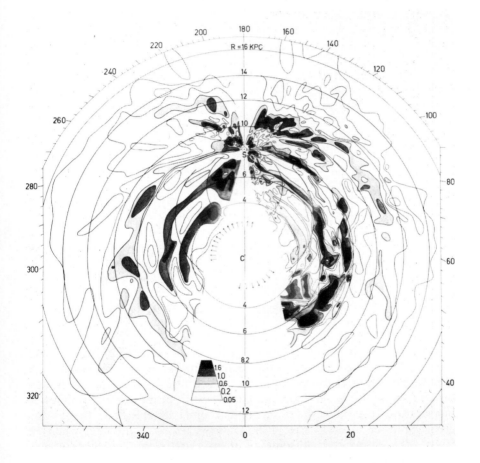

FIG. 7.4.   Radio map at 21 cm wavelength of the distribution of neutral hydrogen based on combined observations at Leiden (Holland) and Sydney (Australia). Distances from the centre are marked in kiloparsecs (1 kpc = 3260 light years). (After Kerr and Westerhout, 1964.)

As we may expect, the gas of interstellar space is very cold, about 100° above absolute zero. Nevertheless, even at this very low temperature the radio emission is easily measurable by sensitive radio receivers. As Fig. 7.4 is a rather complex picture of the distribution of hydrogen, some of the main spiral features are shown in a simplified diagram in Fig. 7.5 (together with central regions not included in Fig. 7.4).

The names for the arms have been chosen rather arbitrarily and are not yet finally decided. For example, the Sun lies in the arm tangentially directed towards the constellation of Cygnus, and hence is often called the Cygnus arm. The Orion Nebula probably lies within this arm, so an alternative name, the Orion arm is sometimes used.

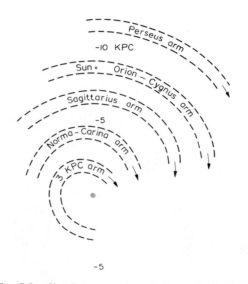

Fig. 7.5. Sketch representing spiral arms in the Galaxy.

Most of the arms have a rotational speed of the order of 200 km/sec. An unusual feature found in one of the central arms, known as the 3 kiloparsec arm because that is its distance from the centre of the Galaxy, is that besides rotating it is also expanding outward from the centre at about 50 km/sec. Does this outward flow mean that sometime in the past there was a great explosion at the centre of the Galaxy? We shall find later when we study other galaxies that some of them show clear evidence of vast explosions at the nucleus.

The 21 cm radio line measurements also reveal that the hydrogen distribution is remarkably thin and flat, at least out to a radius of 20,000 light years. The radio observations have been utilised to define the

standard astronomical position of the plane of the Galaxy and its centre, and to draw up a new system of coordinates for specifying galactic "latitude" and "longitude". The galactic plane is the "equator" and the centre of the Galaxy is set at galactic longitude 0°.

The reliability of the 21 cm line observations of atomic hydrogen as an indicator of the spiral arms where concentrations of gaseous matter condense into stars has been called into question. The regions of star formation are typically sites of molecular concentrations rather than atomic hydrogen, but molecular hydrogen (H_2) has no radio line emission. According to present ideas, the spiral pattern of young hot stars and glowing ionised gas originates from a rotating wave of compression producing a density wave with ensuing condensation of gas and particles into protostars. There is no guarantee that the distribution of atomic hydrogen defines these zones. In fact the evidence suggests that not only is atomic hydrogen more widespread but anomalies in motion cause uncertainties.

Since the discovery in 1970 of the 2.6 mm line of carbon monoxide, an abundant gas in the Galaxy, the line has served as a guide to the general molecular distribution. Most of the carbon monoxide, and the optical spiral arms, are contained well within a radius of 10 kpc from the galactic centre. In contrast atomic hydrogen extends out to at least twice this radius.

It is of interest to note that recent evidence of higher than expected rotation rates in outer parts of the galactic disk and dynamical considerations have led to the inference of a huge corona enveloping the Galaxy out to a radius of perhaps 100 kpc. The corresponding estimates of the total mass of the Galaxy are 5 or 10 times greater than had previously been believed.

Several years ago there seemed hope that the problems of the grand design and evolution of galactic structure would soon be solved. Such a prospect has now receded, and demands a more fully integrated picture of the stellar and gaseous components of the Galaxy.

Continuous Radiation from the Galaxy

The 21 cm hydrogen line constitutes only a small part of the radio power emitted by the Galaxy. Most of the radio emission extends over a wide band of wavelengths and is called the continuous (or continuum) radiation. As we find in considering individual sources in the Galaxy, this kind of emission can arise in two ways; one is thermal radiation from ionised hydrogen, another is by non-thermal processes such as synchrotron radiation from high energy electrons in weak magnetic fields. The problem is to sort them out when both appear together, and measuring the spectrum helps us to do this.

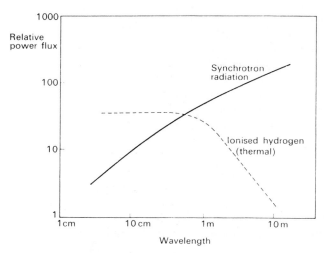

FIG. 7.6.   Spectra of components of galactic radio emission.

Figure 7.6 shows a typical dependence of power flux on wavelengths for the two types of radiation. At short wavelengths thermal radiation predominates, while at long wavelengths synchrotron radiation is by far the stronger. Hence if we map the galactic radiation at wavelengths shorter than 20 cm most of the radio emission arises from ionised hydrogen, which has a temperature of about 10,000 K. The hydrogen has been ionised by hot stars in the galactic disk, and the radiation stands out as a narrow strip, only about 2° wide, extending along the galactic equator. The distribution is clearly similar to that of neutral hydrogen found by the 21 cm line. A connection between ionised and the neutral hydrogen is naturally to be expected. It is believed that the stars are formed from concentrations of hydrogen which become compressed and collapse into stars under gravitational forces. The hot stars produce ultraviolet light and X-rays which ionise surrounding hydrogen. Consequently, ionised gas is found near dense regions of neutral gas.

When maps of the Galaxy are plotted at metre wavelengths, where the radiation is principally synchrotron emission, we see a much wider spatial distribution. A typical map is shown in Fig. 7.7.

The metre-wave distribution can be separated into two parts; a ridge of about 5° width, and a much more widespread emission. The two parts can be distinguished by seeing how the intensity varies in crossing the galactic equator, as shown in Fig. 7.8.

We note that the narrow part of this emission, sometimes called the non-thermal disk, is a few times broader than the hydrogen disk. Extending far beyond these is a very wide distribution, called the non-thermal

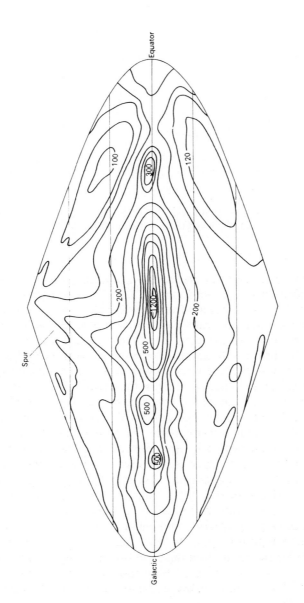

FIG. 7.7.   Radio map of the Galaxy at λ = 1.5m obtained with a 17° beamwidth. The numbers of the contours are radio brightness temperature (°K). The galactic centre lies at the centre of the map. The horizontal lines are marked at intervals of 30° of galactic latitude. (After Droge and Priester, 1956.)

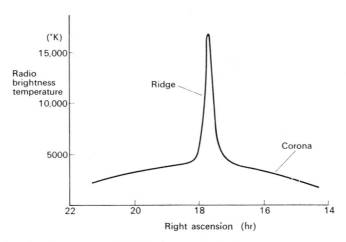

FIG. 7.8.   A section across the Milky Way at λ = 3.5 m indicating the narrow and broad parts of the radio emission. (After Mills, 1955.)

halo or corona. The flux density increases at longer wavelengths (up to λ = 30 m the flux is roughly proportional to $\sqrt{\lambda}$). At wavelengths greater than a few metres, the radio emission from the Galaxy is easily detected with any sensitive radio receiver, and can be heard as a hissing noise on a loudspeaker. Some noise may come from the receiver, but galactic radiation will appear as an increase of noise when the aerial is connected.

Most of this non-thermal radio power comes from galactic space where there is very little hydrogen. The strength of the radiation and its spectrum shows without doubt that it must originate by the synchrotron process. The essential ingredients for this are very fast electrons, and a magnetic field to cause deviations in the electron path. Only a very low density of electrons and a very weak magnetic field are required. We know that there are high energy particles in the form of cosmic rays, and we can infer that very fast electrons with speeds approaching that of light are present. We know too that there is optical evidence of weak magnetic fields of the order of a millionth of a gauss (or 0.01 tesla, T).

The explanation of galactic radio emission as synchrotron radiation was first suggested by Kiepenheuer in 1950, the year that Alfvén and Herlofson made a similar proposal to explain the origin of radio emission from the non-thermal discrete sources. Subsequent research confirmed these hypotheses. There is a striking similarity between the general radio emission from the Galaxy and that from localised intense sources in the Galaxy, such as the supernova remnants which will be discussed in Chapter 8. Both are characterised by strong radio emission and a spectrum given by power flux proportional to λ^x, where x lies in the range 0.3 to 0.8. We

explain both in terms of electron energies of the order of 10^9 eV and magnetic fields of between a thousandth and a millionth of a gauss. This raises an interesting question—could the total galactic background be due to the combined effects of many old supernova remnants, some by now greatly expanded and intermingling to fill galactic space? Was there also perhaps a far more intense explosion in the galactic nucleus? These are indeed possibilities. As we shall see later in Chapter 9, there is evidence of extraordinarily violent explosions taking place during the evolution of certain galaxies. We will here mention some features of the distribution and polarisation of the radiation from the Galaxy which may help to throw some light on the origins of galactic radio emission.

It is now generally accepted that the continuum radio emission is mostly synchrotron emission from high energy electrons in the weak magnetic fields permeating galactic space. As the synchrotron process produces linearly polarised radiation, one might expect that polarisation would easily be observed in galactic emission. For many years no linear polarisation was detected, and it was assumed that tangled magnetic fields and Faraday rotation had so effectively mixed the initial polarisations as to make the received signals appear randomly polarised. However, it has been known since 1949 that in some directions starlight is polarised, and the most plausible explanation is that the optical polarisation is caused by scattering from elongated dust particles aligned by interstellar magnetic fields of strength assessed in about 0.1 tesla. The optical observations indicated a nearly uniform field running across one region of the sky. It was not until 1962 that the expected linear polarisation associated with synchrotron radio emission was detected; probably it had been missed before because of its patchy distribution, and the low percentage polarisation due to the depolarising influence of Faraday rotation. The Dutch radio astronomers were the first to be successful in measuring linearly polarised emission from the galactic continuum. The radio and optical evidence of the strength and direction of the magnetic field is in good agreement. The areas of linearly polarised radio emission are almost all contained in a band round the sky. This can be explained on the synchrotron theory as radiation from the spiral arm in which we are situated, with the magnetic field along the arm. The polarisation of the local spiral arm stands out because it is nearest to us, and has a simple field structure. The recognition of polarised radiation gives good confirmation that the radio emission is produced by the synchrotron process.

Galactic Spurs and Ridges

The galactic radiation is by no means a smooth distribution but shows detailed structure as well as some large-scale features. One of the most prominent is the North Polar Spur which can clearly be seen in the radio

map of the Galaxy shown in Fig. 7.7. In the southern galactic hemisphere there is another ridge, called the Cetus arc because it passes through the constellation of Cetus. These ridges or spurs lie along large circular arcs in the sky giving the impression that they may be shells of old supernovae. This seems quite a plausible explanation although the absence of optical emission indicates that if they are supernova remnants their optical luminosity must be very low.

The Galactic Nucleus

Since the earliest observations by Jansky and Reber of the continuum radio noise it has been known that the greatest intensity is concentrated toward the centre of the Galaxy. We may well ask, what is there at the centre, at the nucleus of our Galaxy? The heavily obscuring dust clouds mask the optical view. If we look at other spiral galaxies resembling our own we find that they have very bright central regions, presumably containing densely packed stars.

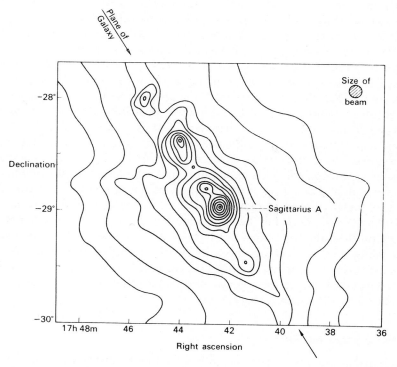

Fig. 7.9.   Map showing the region of the galactic centre observed at λ = 10 cm. (After Cooper and Price, 1964.)

As we cannot see the centre of our Galaxy we have to rely on radio and infrared evidence of the conditions at the nucleus. The direction of the centre lies in the constellation of Sagittarius, and a radio map of the region is shown in Fig. 7.9. At the centre is an extremely compact intense radio and infrared source. VLBI measurements have indicated a source size of no more than 10 A.U. The high-velocity motions detected in dense compact clouds of ionised gas in the surrounding inner zone of the Galaxy suggest a supermassive central object possibly equivalent to 50 million solar masses.

With the help of radio astronomy we have now a much clearer picture of our Galaxy. To discover more about the evolution of galaxies we must look at many other galaxies of various ages in the universe.

8. Radio Sources in the Galaxy

WHEN we look around the sky on a clear night, almost all the bright points of light we see are stars in our Galaxy, and in directions where they are numerous and distant they merge into a band of light known as the Milky Way. Amongst the stars, two nebulous sources can just be discerned with the naked eye. They have a luminous, misty appearance and hence the name "nebula" (Latin for "cloud"). One is the Andromeda Nebula, a separate galaxy like our own, at a distance of more than two million light years. The other nebula visible to the naked eye is quite a different kind and closer to us. This is the Orion Nebula, a cloud of luminous gas about 1500 light years away and situated within our own Galaxy.

If we look at the sky through an astronomical telescope many more stars come into view, and more nebulae are revealed. In 1784 the French astronomer Messier, impressed by their hazy, non-stellar appearance, made the first catalogue of nebulae. The Andromeda Nebula is M 31 (that is, number 31 in Messier's list) and the Orion Nebula is M 42. Messier had not the knowledge of the distances and nature of the nebulae to be able to distinguish the different types. Later, when distances could be estimated, it was realised that they could be divided into two main groups, the external galaxies and the visible clouds of gas in our Galaxy. In this chapter we shall be concerned with sources in the Galaxy where the objects that can be seen comprise vast numbers of stars and the visible clouds known as galactic nebulae.

Thermal Emission Nebulae in the Galaxy

For many years the only objects within the Galaxy corresponding to radio sources were found to be nebulae. One important class comprises the diffuse emission nebulae, which show characteristic emission lines in their optical spectra. They consist predominantly of vast regions of ionised hydrogen surrounding very hot stars of high luminosity, and the Orion Nebula is the best known example of this type. Their radio emission is a classic example of thermal radio waves originating from the hot electrons of ionised hydrogen.

The ionisation is caused by ultraviolet radiation from very hot, large stars. What happens is that the ultraviolet radiation ejects electrons from the hydrogen atoms of surrounding interstellar gas. When an electron recombines it cascades down through different energy levels of the

hydrogen atom causing the emission of light. Very hot, large stars, comprising a class designated as O type, have temperatures ranging from 30,000 to 70,000 deg. In the next category are the B type stars often found fairly close to O stars. The occurrence of hot stars of high luminosity in the vicinity of comparatively dense regions of hydrogen gas is no mere chance. It is here that the stars are actually being formed from the hydrogen, the gas which occurs so extensively throughout space. In the vicinity of the denser concentrations of hydrogen there are usually large quantities of dust

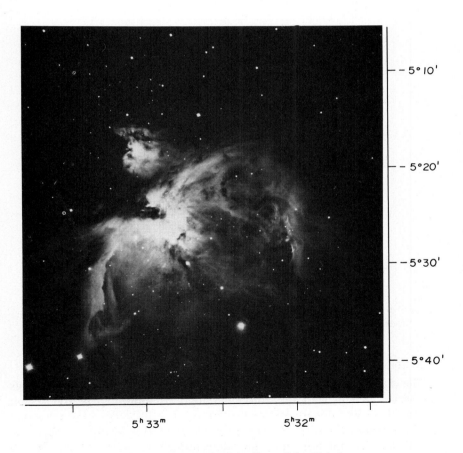

FIG. 8.1. (a) Photograph of the Great Orion Nebula.

particles, and the force of gravity brings the gas and dust together to form new stars. Large masses become highly compressed and so hot that nuclear reactions occur producing O and B stars of very high temperature. O and B classes comprise new stars in a very active phase lasting for about a million years, by which time they have lost so much energy that they have evolved into cooler types of stars.

In the Orion constellation there is a vast complex of hydrogen and hot stars, although much is obscured by dust. Some of the dust clouds are very dense and well-defined forming the so-called dark nebulae, looking like storm clouds against a bright sky. One such nebula in the Orion constellation is the Horsehead Nebula, with its striking resemblance to the head of a black horse. The brightest emission nebula in the Orion complex is the well-known Orion Nebula, M 42, and a photograph is shown in Fig. 8.1(a)

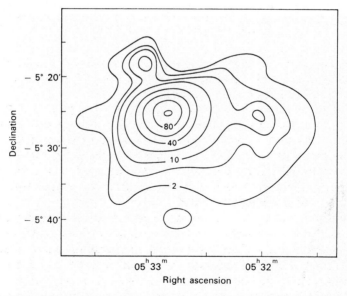

FIG. 8.1. (b) Radio map of the Orion Nebula at λ = 73 cm. The numbers on the contours are in units of 66°K radio brightness temperature. (After Mills and Shaver, 1967.)

The diameter of the Orion Nebula is almost ½°, and Fig. 8.1 compares the visual nebula with the adjoining radio contour map of the region obtained at a wavelength of 73 cm with the new Mills Cross radio telescope giving a beamwidth of 3′. The ionisation of the Orion Nebula is caused by the ultraviolet radiation from the Trapezium cluster of bright stars at the centre, containing two O stars and several B stars.

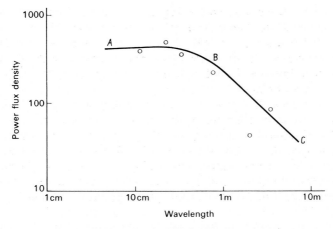

Fɪɢ. 8.2.   Radio spectrum of the Orion Nebula showing measured values (circles) compared with a calculated curve for an electron temperature of 10,000°K. The units of power flux density are Jy. (after Menon, 1964.)

The radio spectrum of the nebula, that is, the way the radio emission depends on wavelength, tells us a great deal about the temperature and density. The spectrum is shown in Fig. 8.2.

We explained on page 15 that at very long wavelengths an ionised gas becomes opaque to radio waves, and radiates simply as a hot body according to its temperature. The intensity of radiation is given by the Rayleigh–Jeans Law, $P = kT/\lambda^2$. This corresponds to the part BC of the spectral curve shown in Fig. 8.2. If the intensity is determined by measurement then we know at once the electron temperature of the ionised gas, which is about 10,000 K.

At shorter wavelengths, the ionised gas is more transparent to radio waves and $P = \epsilon kT/\lambda^2$, where the factor ϵ is proportional to λ^2, as well as depending on the total electron content N in the nebula and the temperature T. The wavelength terms now cancel out so that the power received from the source is practically constant as shown in the part AB of the spectral curve, depending only on N and T. As we know the temperature T, the power radiated now indicates the number of electrons. It is found that the electron density at the centre reaches a few thousand electrons per cubic centimetre, diminishing to about 10 towards the outer parts of the nebula.

All this information fits well with optical observations of the nebula. The radio information on nebulae is interesting for two reasons. The first is that they provide excellent examples of thermal radio emission from a hot ionised gas. More important for astronomy, radio waves go through the obscuring dust so that only by radio can we study certain nebulae where the interstellar fog makes them difficult or impossible to see.

Several radio lines have been observed corresponding to electron transitions in outer excited levels of hydrogen in the vicinity of ionised hydrogen (H II regions) surrounding very hot stars. Only transitions to an adjacent outer level have the appropriate energy change to produce radio waves. The first radio detection of this type was made by the Russians in 1964 of line emission from the Orion Nebula corresponding to an electron transition from the 105th to 104th level in atomic hydrogen. Since then many other lines have been detected in hydrogen as well as some in helium, and other atoms. These lines are known as recombination lines since they are a consequence of electrons recombining with positive ions. The electrons falling into higher atomic levels cascade down through a series of transitions to lower levels so emitting line radiation in the process. Examination of radio recombination lines gives much additional information on H II regions, in particular velocities and temperatures. Combined with date on the continuum emission a remarkably complete picture of H II emission in obscured regions of the sky can be obtained.

Supernova Remnants

THE CRAB NEBULA

An entirely different kind of nebula can be found in the Galaxy, and by far the most striking example of this class is the Crab Nebula, so called because its shape has some resemblance to a crab. A photograph is shown in Fig. 8.3. Messier was impressed by its unique appearance, and he put it at the top of his list, so it is designated M 1. The Crab Nebula is now known to be the remnant of a supernova, a gigantic explosion of a star.

A stellar outburst, indicated by a sudden great increase in the brightness of a star, is known as a nova. A fairly recent example of a bright nova in our Galaxy occurred in 1918 when a very faint star (magnitude 11) suddenly brightened by 70,000 times and for a few hours outshone every other star in the sky. In the following months the gas from this explosion could be observed expanding into a nebulous shell as the nova gradually faded in brightness. On rare occasions a far more intense explosion known as a supernova may occur, with an increase of brightness of perhaps 100 million times.

Chinese chronicles provide the most comprehensive source of information on supernova outbursts. In ancient China, systematic observations were made of unusual celestial phenomena, and the annals of each dynasty included a section devoted to notable astronomical events because it was believed that such events had a profound effect on the course of history. There it is recorded that in 1054 a supernova, or "guest star" as the Chinese called it, occurred in a position within the constellation of Taurus.

It is not only the agreement in position that identifies the 1054 supernova

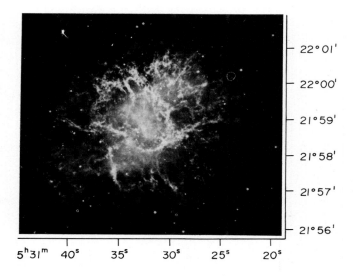

FIG. 8.3. (a) Photograph of the Crab Nebula.

with the Crab Nebula. Optical examination of the remnants enables us to establish in a most striking way the approximate date of origin of the Crab Nebula. Optical line spectra have Doppler shifts indicating that the front is approaching at about 1100 km/sec and the back receding at the same rate. The nebula is obviously expanding at tremendous speed. Photographs taken several years apart confirm this rapid expansion because the angular width of the nebula is increasing at a rate of about $1/5$ sec of arc per annum. From these rates and the present size we can find when the expansion began, and in this way we deduce that the Crab Nebula originated from a vast stellar explosion about 1000 years ago. The combined evidence of date and position leaves no doubt that the Crab Nebula must be the remnant of the 1054 supernova. The distance calculated from the expansion rates is approximately 5000 light years, agreeing well with other estimates.

The Crab Nebula is a strong radio source, and it was in fact the first discrete radio source to be successfully identified with a visible object. In 1948 Bolton and his colleagues in Australia determined the position of the

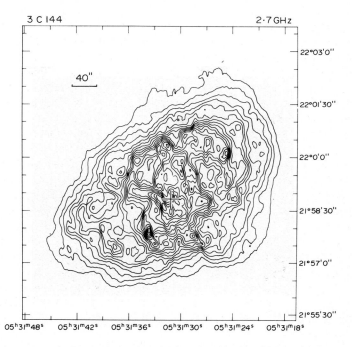

FIG. 8.3.(b)   Radio map of the Crab Nebula at λ = 11 cm. The contour interval is 250 K in radio brightness temperature. The cross marks the position of the pulsar. (After Swinbank and Pooley, 1979.)

radio source Taurus A very accurately with an interferometer, and found that it concided with the Crab Nebula. With the improvements in resolution now available it has been possible to map the radio distribution as shown in Fig. 8.3(b).

The Crab Nebula is a most remarkable optical and radio source, and it provided the first positive evidence of the main cause of the powerful non-thermal radio emission from astronomical sources. In 1950 two Swedish physicists, Alvén and Herlofson had suggested that the strong radiation from radio sources might arise from electrons with high energies like those of cosmic rays undergoing accelerations due to the action of magnetic fields, the process known as synchrotron radiation discussed previously in Chapters 2 and 7. One of the expected characteristics of this type of emission is that it should be polarised, with the radiated electric field at right angles to the magnetic field of the source. Unfortunately, tangled effects of magnetic fields existing in the ionised gas along the path of the radiation on its way to the observer can so twist the radio polarisation that only a small percentage of the polarisation remains to be

distinguished by the observer at the Earth. Depolarisation affects the longer waves most of all. Now in 1953 it was discovered that the optical radiation from the Crab is strongly polarised. This led the Russian astronomer Shklovsky to suggest that the continuous optical radiation as well as the radio emission might arise from the synchrotron process. As would be expected, the optical emission shows a strong linear polarisation because at short wavelengths depolarisation has little effect.

Confirmation that the radio waves originate by the synchrotron process came in 1957 when Mayer and his colleagues at the Naval Research Laboratory, USA, demonstrated that the radio emission also shows linear polarisation. They found 7 per cent mean polarisation, the low value easily being attributable to depolarising effects. This was the first observation of linear polarisation in a discrete radio source, and it left no further doubts on the validity of the synchrotron interpretation.

The Crab Nebula was also the first galactic object to be identified as an X-ray source during observations made in 1964 from a rocket in the USA by Friedman and his group at NRL. The angular extent was found to be about 1', appreciably less than the radio or optical size of the nebula. Later observations revealed about 15 per cent linear polarisation indicating that the X-ray emission like optical and radio is also produced by the synchrotron process.

The shape and continuity of the power spectrum from radio and optical to X-rays and γ-rays adds a further convincing argument that the same basic process operates throughout this enormous electromagnetic waveband from at least 10^6 to 10^{20} Hz. The spectrum is illustrated in Fig. 8.4.

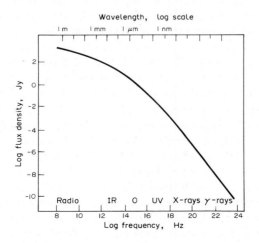

FIG. 8.4.   Spectrum of the Crab Nebula (from γ-ray to radio wavelengths).

The spectral index (indicated by the slope of the spectrum) falls at the longer wavelengths from ~1.2 for X-rays, ~0.9 for optical to ~0.25 for radio. We infer that the radiation arises in a magnetic field of around 0.5 nT from electrons with energies of the order of 10^{14} eV for X-rays, 10^{11} eV for optical and 10^8 eV for radio waves. Although the spectrum shows that the power flux per unit bandwidth is less at optical and lower still at X-ray frequencies, in comparison with radio they occupy greater bandwidths. Consequently, the integrated power is much the same in all three bands, approximately 10^{30} W.

The question immediately arises, how do the fast electrons, particularly those producing optical radiation and X-rays, sustain their energy unless there is continual replenishment? The lifetime of the energetic electrons emitting radio waves could be 1000 years, but the optically radiating electrons lose their energy more rapidly with a lifetime less than 100 years, while for the highest energy X-ray electrons the lifetime is only ~2 years. It is now surmised that the nebula is energised not only by the ejected matter blasted off from the supernova explosion, but also by the continuing transfer of energy from the residual core of the exploded star. Although many aspects of the phenomena remain speculative, the discovery of the pulsar at the core has provided the clue to the replenishment of the energetic electrons and fields to maintain the synchrotron radiation. The Crab pulsar, the "engine" sustaining the output of radiated power, is discussed later (p. 160) in the section on pulsars.

A few more details should be mentioned to complete the brief picture of this fascinating supernova remnant which it may be noted is variously designated M 1 or NGC 1952 (optical nebula); SNR 1054 (supernova remnant); Taurus A or 3C 144 (radio); Taurus X-1 (X-rays). The reduced size of the optical nebula compared with radio and the still smaller X-ray source accords with the higher loss rate of the more energetic electrons. The filaments seen in the light of the Hα emission line (Fig. 8.3(a)) contain two or three solar masses of ejected matter from the supernova explosion and comprise gas ionised by the UV and higher energy synchrotron radiation. Currents flowing along the ionised filaments are probably responsible for the higher magnetic fields pervading parts of the nebula, which would account for associated ridges in the map of radio emission. The total mass of the particles generating synchrotron radiation (which includes the amorphous optical continuum) is almost negligible; their energy derives from their enormous relativistic speeds.

The mass of the pulsar is estimated at ~1.5 times that of the Sun. A very faint elliptical halo, about 6′ by 14′ in extent, containing roughly 4 solar masses of gas, surrounds the nebula. Hence the total stellar mass of the original presupernova star must have been about 8 solar masses. Supernovae of this kind occurring in large newly formed stars in dense parts of the galactic disk are known as Type II. In certain respects, however, the

Crab supernova remnant possesses unique properties. No other known remnant emits continuum synchrotron radiation over the entire electromagnetic spectrum. The amorphous shape of the Crab Nebula is rarely found, and the common feature of most supernova remnants, is a shell-like structure, particularly evident in radio maps.

TYCHO BRAHE AND KEPLER'S SUPERNOVAE

Supernovae are rare phenomena and it has been estimated that their rate of occurrence in the Galaxy averages about one per 30 years. How easily or not they may be seen depends on their distance, and whether they appear in clear directions in the sky. If a supernova explosion occurred in a star at 30 light years distance from us it would for a time turn night into day. Supernovae occur at irregular intervals and positions, and it is a remarkable chance that two easily visible supernovae were recorded within an interval of 32 years. The first, in 1572, was carefully observed by the Danish astronomer Tycho Brahe. In its early stages it outshone Venus and was visible in broad daylight. The second in 1604, rather less bright, was recorded by the German astronomer Kepler. The optical remnants of these supernovae are now very faint. Some wisps and filaments from Kepler's supernova were successfully photographed in 1943 by the 2.5 m Mt. Wilson telescope. These remnants have low intrinsic brightness, and interstellar dust further weakens their visibility. The remnants of Tycho Brahe's supernova have proved even less conspicuous, but have been photographed with the aid of the 5 m Mt. Palomar telescope.

Both these supernovae display much similarity and belong to the same general class known as Type I. Supernovae of this type are explosions of fairly old stars with mass not greatly exceeding that of the Sun. They can be recognised by their characteristic rise and fall of initial brightness.

Although the optical remnants of supernova such as Tycho Brahe's and Kepler's are delineated by faint filaments, much more outstanding are the prominent shells detectable both at radio and X-ray wavelengths. Although closely associated in structure different radiation mechanisms are involved. A radio map and X-ray picture of Kepler's supernova are shown in Figs. 8.5(a) and (b). The radio flux from Tycho's supernova remnant is stronger and its angular size larger (7') than Kepler's (3') mainly because the Tycho remnant is nearer. Their distances are known only approximately, the Tycho remnant at 5 kpc ($\sim$15,000 light years) and Kepler's at about twice this distance. For both supernova remnants the mean velocity of expansion is estimated at 12,000 km/sec.

It is clear that the shell structure represents the effect of an expanding shock wave originating from the supernova explosion. The shock front spreads outwards at supersonic speed compressing and heating the interstellar gas. The patchy pattern can be attributed to the irregular cloud

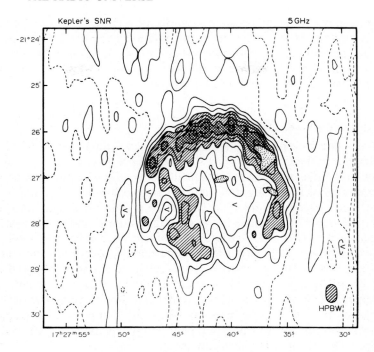

FIG. 8.5.(a)   Radio map of Kepler's supernova remnant $\lambda = 6$ cm.
The contour interval is 2 K. (After Gull, 1975.)

distribution of interstellar gas and to imperfect symmetry of the initial
explosion. The X-ray spectrum corresponds to thermal radiation from hot
ionised gas, and the X-ray spectral lines indicate that the passage of the
shock wave has raised the temperature to the order of 10 million deg K.
The radio emission is generated by a different process, for the spectrum
(spectral index ~0.7), the intensity, and the polarisation all poi..t with
certainty to the synchrotron mechanism. The spherical uniformity in the
direction and distribution of linear polarisation in Tycho's supernova
remnant shown in Fig. 8.6(b) is outstanding. The shock wave evidently
creates the appropriate conditions for synchrotron radio emission. Its
passage compresses the interstellar magnetic field, whilst relativistic electrons are already present (as cosmic rays) in the interstellar medium and
more are likely to be accelerated within the remnant both in the explosion
and in the expanding shock and ensuing turbulence.

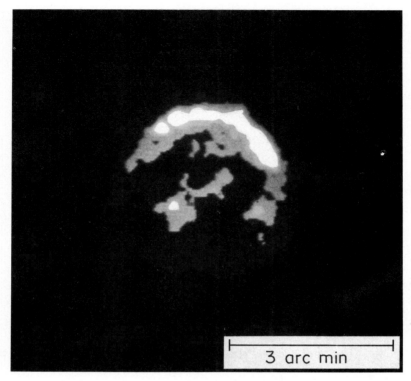

3 arc min

FIG. 8.5.(b)   X-ray image of Kepler's supernova remnant obtained with the high resolution images on the Einstein Observatory satellite. (After Helfand, 1980.)

CASSIOPEIA A

The strongest of all the radio sources in the sky, discovered in 1947 in the constellation of Cassiopeia, was designated Cassiopeia A. With such a powerful source it was puzzling that no obvious identification with any bright star or nebula could be found. It was realised that any visible object corresponding with the radio source must be very faint, so the next step was to determine the radio position more precisely. There are so many faint visible objects in the field of view of a powerful telescope that it is only possible to demonstrate that a radio source coincides with one of them if the radio position is known very accurately. In 1951, F.G. Smith at Cambridge succeeded in measuring the radio direction by means of interferometers to an accuracy of 10″ in Right Ascension and 40″ in Declination. This accurate position enabled Dewhirst of the Cambridge astronomical observatory to point out that photographs showed a faint

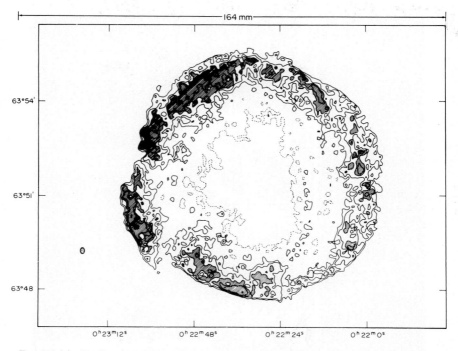

FIG. 8.6.(a)   Radio map of Tycho Brahe's supernova remnant at $\lambda = 6$ cm. (After Duin and Strom, 1975.)

nebulosity closely coinciding with the radio source. Then Baade and Minkowski in America obtained more detailed pictures of the nebula with the 5 m Mt. Palomar telescope which showed fragments of a circular shell with an outer radius of about 2′ agreeing well with the radio position. The nebulosity contains many small condensations, and photographs taken at different times show that most of them are moving outward from the centre at high speed. The high speed of expansion is also evident from the Doppler shifts of the spectral lines indicating an expansion velocity of about 7500 km/sec. This speed, combined with the apparent movements of the nebulosities, tells us what the distance must be (in just the same way that the apparent rate of movement of an aeroplane across the sky depends on its distance). In this way Cassiopeia A is estimated to be about 10,000 light years away. The same distance is indicated by observations of absorption by the 21 cm hydrogen line. There are three spiral arms in the direction of Cas A but only two 21 cm line absorption dips appear in the spectrum. This firmly locates the source between the second and third arms and confirms the distance at about 10,000 light years.

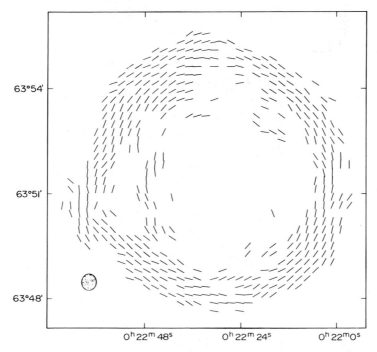

FIG. 8.6.(b)   Polarisation map (electric field) of Tycho Brahe's supernova remnant derived
from radio observations. (After Duin and Strom, 1975.)

From the rate of expansion of fast-moving optical features the explosion date has been estimated at about 1660, making the age of the nebula about 320 years, the youngest known supernova remnant in the Galaxy. The mass of the expelled matter inferred from the visible fragments is comparable to the mass of the Sun. The supernova star must have been a massive object of at leat 10 solar masses producing a Type II supernova. Why was this supernova not observed visually and recorded by astronomers? Observatories were already established in Europe, and Tycho's and Kepler's star explosions had been clearly seen and studied almost a century earlier. The most likely explanation is obscuration by interstellar dust in the direction of the central region where the stellar explosion occurred. It has been suggested that much of the obscuration may have been caused by a rapidly formed dust cloud developed by the supernova explosion itself.

The Cas A supernova remnant appears as an irregular shell of about 5′ diameter at radio and X-ray wavelengths. A contour diagram of the radio intensity is illustrated in Fig. 8.7. As in the Kepler and Tycho remnants different processes are responsible for the radio and X-ray emission

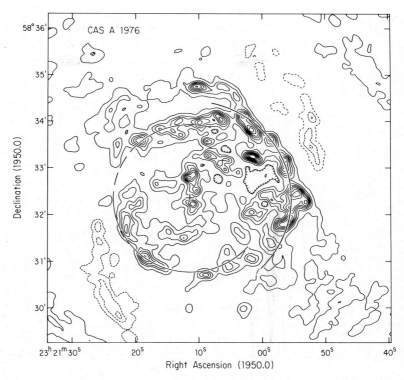

FIG. 8.7.   Radio contour map of Cassiopeia A at λ = 11 cm. (After Dickel and Greisen, 1979.)

although both are stimulated by the expanding shock. The radio spectral index (~0.8) and high intensity indicate synchrotron radiation, whilst the X-rays arise from heated gas at a temperature of around 10 million deg K.

The irregular form of the shell undoubtedly reflects the variations in density of the interstellar gas on which the shock wave impinges. Detailed examination of optical, radio and X-ray distributions within the remnant reveal a complexity of compact features. Few of them coincide in the different radiation bands and many show marked changes in time scales of a few years. Most compact radio knots, for instance, exhibit seemingly random velocities and variations of radio brightness, with an average outward trend and decrease of total power. Figure 8.8 shows a remarkable photographic representation of radio intensity derived from observations at λ = 6 cm with the Cambridge 5 km radio telescope. The variability in detailed structure indicates considerable turbulence. As the supernova expands against the surrounding interstellar medium it will eventually lose

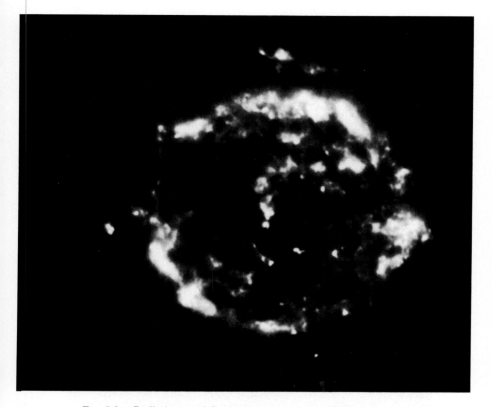

FIG. 8.8.   Radio image of Cassiopeia at $\lambda = 6$ cm. (After Bell, 1977.)

its momentum. The energy of the field and of the particles must gradually subside. The Russian astronomer Shklovsky made theoretical calculations of the rate of decline of Cassiopeia A and predicted that the radio emission should weaken by 1 or 2 per cent per annum. Careful measurements of the flux density were then compared with those obtained in earlier years and it was clearly established that the radio emission is decreasing by about 1 per cent per annum, a remarkably good verification of Shklovsky's prediction.

The observed decline of radio power is not precisely the same at all wavelengths. In fact measurements at $\lambda \sim 8$ m revealed that during the 1970s a surge of around 40 per cent in radiated power occurred with a duration of a few years. There are also indications of the existence at metre wavelengths of a radio spur on the eastern side (left in Fig. 8.7) apparently associated with a jet formed by fast optical knots breaking through the main shell. However, no detailed maps at long wavelengths are as yet available.

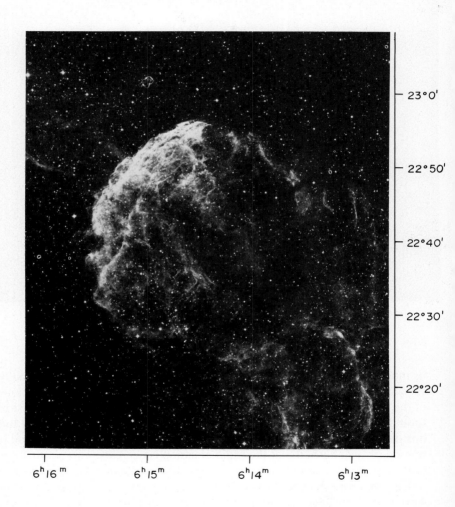

FIG. 8.9.(a)   Photograph of the supernova remnant IC 443.

SUPERNOVA REMNANT IC 443

As an example of a very old supernova remnant we shall discuss the nebula IC 443 in the constellation Gemini and shown in the photograph, Fig. 8.9(a). The irregular shell at ~ 5000 light years distance possesses a particularly striking almost circular segment of about 45' diameter. The expansion velocity is 65 km/sec, and this Type II supernova remnant with an estimated age of 60,000 years is in a very late stage of expansion.

The radio map in Fig. 8.9(b) shows a close correlation with the distribution of optical filaments. In contrast, other supernova remnants such as Cas A do not exhibit such detailed correspondence. The optical line emission of IC 443 is characteristic of ionised gas at a temperature of

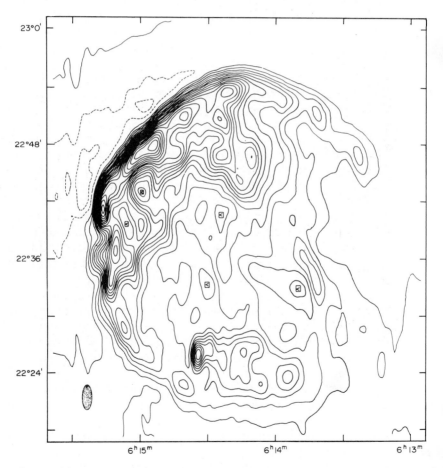

FIG. 8.9.(b)   Radio map of IC 443 at λ = 21 cm. (After Duin and van der Laan, 1975.)

10,000 K and an electron density of $\sim$250 cm^{-3}. The radio emission, however, is typically synchrotron radiation in its spectral index ($\sim$0.5) and partial linear polarisation. Explanation must therefore be sought for the very close association between radio and optical features. One interpretation is that the filaments after being heated by the shock to more than 10,000 K subsequently cool and contract so compressing the magnetic field. Relativistic electrons already present then give rise to the radio emission.

A particularly interesting possibility is that the nearby pulsar PSR 0611 + 22 might be the residual star of the supernova explosion. At 0°.7 from the centre the pulsar is located just outside the remnant. The pulse period, about 1/3 sec, and rate of change $\sim$0.5 nsec/day, suggests a pulsar age of around 65,000 years, similar to that of the remnant. To reach its present position the pulsar must have been moving at over 100 km/sec, and this is comparable with known pulsar velocities.

THE VELA SUPERNOVA REMNANT

Vela-X is one of the largest and brightest supernova remnants in the Southern Hemisphere. Optically, the Vela remnant, approximately 1500 light years distant, is seen as a complex mosaic of filaments and nebulosities covering 4° by 2°. The light is principally emission from ionised hydrogen at a temperature of the order of 10,000 K. The radio structure is irregular also, and takes the form of a filled distribution rather than a shell. Linear polarisation and the spectral index ($\sim$0.3) indicate synchrotron radio emission. A short-period pulsar discovered close to the centre is without doubt the residual star. These points of resemblance to the Crab Nebula are most interesting, and it seems probable that as in the Crab the pulsar is continuing to feed energy to the amorphous radio remnant. However, there is no evidence to suggest that either the optical or diffuse X-ray emission associated with the Vela supernova represent synchrotron radiation and in detail the structures are markedly different.

It is indeed possible that a clearly observable supernova might occur in our Galaxy at any time, and such an event would undoubtedly command the enthralled attention of astronomers throughout the world. It seems extraordinary that our knowledge of supernovae prior to the sixteenth century is contained only in the ancient records of China and certain Oriental countries where the rulers were concerned with the implications of the potents of unusual celestial phenomena on the future of their dynasties.

In concluding our discussion of the physical processes involved, another far-reaching consequence of supernovae must be mentioned. In the process of compression and heating to extremely high temperatures during the collapse and subsequent explosion of stars a great variety of elements

are formed by nuclear fusion. It is fascinating that the heavier elements have been synthesised in this way and subsequently thrown out by the stellar disruption into interstellar space to become part of the gas condensing again into new stars and in some instances associated planetary systems. All the terrestrial heavy elements we know, including precious metals, have in past ages been involved in this amazing sequence.

In the next section we shall discuss other remarkable and unusual types of stellar activity, and the influence of highly condensed stellar objects including black holes on surrounding material and nearby stars. There must be many supernova remnants in the Galaxy. Several are known with some precision, but others must have merged into the general background of galactic radiation. It is in fact a possibility that much if not all the non-thermal radiation from the Galaxy has originated from supernova explosions, which may well have been the source of the high energy protons and electrons and magnetic fields that permeate the Galaxy.

Radio Stars

FLARE STARS

For almost two decades following the first observation of radio waves from the Sun, no other star had been positively identified as a radio source. It was realised that if stars radiated no more powerfully than the Sun they would be unlikely to be detected. However, it was appreciated that some stars might produce flares far more intense than those on the Sun. The known sporadic increases in the optical brightness of certain red dwarf stars (by as much as several magnitudes) encouraged Lovell at Jodrell Bank in the 1960s to search for corresponding outbursts of radio emission. UV Ceti is usually regarded as the prototype of flaring red dwarfs, often called UV flare stars. Assisted by optical watches organised by the Smithsonian Observatory, USA, and the Crimean Observatory, USSR, Lovell succeeded in establishing coincidences between optical flares and radio bursts for several stars. Confirmation of the stellar origin of the radio bursts was later demonstrated by radio interferometry, and radio brightness temperatures up to 10^{15} K were estimated. The stellar radio flares, like the major Type II outbursts on the Sun, are largely confined to the longer wavelengths ($\sim$ metres). It should perhaps be noted that most of these stars are members of binary systems with separations between the components of typically 1000 times the stellar radius. Subsequently, a much wider range of flaring radio stars has been revealed with their radio emission predominantly at short wavelengths ($\sim$ cm, mm). We shall now discuss the various types which are generally associated with close binary systems.

BINARY STARS

Transient activity of variable radio stars presents complex phenomena, and it is hardly surprising to find some uncertainties of interpretation. Binary systems also provide the right conditions for interactions, and this is especially true when the stars are close together. Consequently binaries figure prominently as radio sources. Possible causes of stellar radio emission are therefore star-spots and flares, and the interaction of stellar atmosphere in close binaries.

Certain red supergiants flare sporadically, an activity which is thought to be correlated with large star-spots. The radio flaring of Antares A presents one example. But there is also a steadier source closer to its companion Antares B, believed to be a shock-excited region of interaction between the stellar winds from the two components.

Algol (β Persei), because it has been extensively observed, is often regarded as the prototype of radio binaries. It has been particularly notable for its high level of radio flaring with quiescent intervals of weeks or months between bursts of activity. Actually, Algol is a triple system but two of the stars constitute a close binary. VLBI measurements during certain events established a source size of the magnitude of the space between the two stars of the close double.

Attention has been drawn in recent years to a class of stars typified by the binary RS Canum Venaticorum, so similar stars are now called RSCVn binaries. They comprise two stars, one about 1000 K hotter than the Sun and the other correspondingly cooler. The light curve, one star partially eclipsing the other with a period of several days, shows a wave attributable to huge star-spots. X-ray emission indicates very high coronal temperature whilst powerful centimetric radio bursts seem analogous to Type IV bursts on the Sun.

It is evident that close binary systems can lead to complicated phenomena, so the introduction of a biological term may not seem extraordinary. One class of binaries is known as the symbiotic stars, the word "symbiosis" originating from a term defining a mutual partnership of two different organisms. The symbiotic stars present a combination of a red giant with an extensive atmosphere at a few thousand deg K and a small compact companion usually a white dwarf. Occasionally, symbiotic stars exhibit intense optical brightening followed by a gradual return to normal intensity within perhaps a few months. The likely explanation is that spillover of gas from the extended atmosphere onto the white dwarf produces intense heating sufficient to induce explosive nuclear reactions. Several symbiotic stars produce radio emission, probably a thermal process.

Before proceeding further we must note the evolutionary sequence of a normal star. In a star like the Sun compression and consequent high temperatures at the interior make hydrogen nuclei fuse into helium thus

releasing thermonuclear energy. The resulting radiation pressure halts the gravitational contraction of the star which settles into a relatively stable state where it spends most of its lifetime. When the hydrogen has been consumed, star contraction is resumed producing more intense heating of the core where helium nuclei then fuse together to form carbon, oxygen and other elements up to the atomic mass of iron. The release of energy causes the outer regions of the star to expand until it becomes a red giant. Finally, with the exhaustion of nuclear burning in the core, the star gradually collapses into a white dwarf of size similar to the Earth. In this final state of degenerate matter the electron shells around the atoms have broken down and the nuclei and electrons are so closely crammed that the density is several thousand tonnes per cm^3. It is believed that 999 stars out of every 1000 are destined to become white dwarfs. What is the ultimate fate of more massive stars? We shall explain in the sections on pulsars and X-ray stars, how radio and X-ray astronomy have led the way in demonstrating the existence of even more highly compressed states. For the moment, however, we shall pursue further the mechanisms of activity and mass transfer between the stars of binary systems.

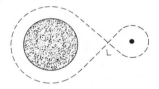

FIG. 8.10.   Illustration of the Roche Lobe surrounding close binary stars.

The interchange of matter and consequent transfer of energy from one star of a close binary system to another can lead to varied and dramatic phenomena. We may picture an imaginary surface around the stars beyond which the gravitational attraction of a component can no longer retain its atmosphere. The shape of the surface, called the Roche Lobe, is illustrated in Fig. 8.10. A close binary system consisting of a large star such as a red giant together with a compact star such as a white dwarf provide ideal conditions for atmospheric flow from the giant star. If the atmosphere of the larger star more than fills the Roche Lobe, then gas in the neighbourhood of the intermediate position L, called the Lagrangian point, can stream into the sphere of attraction of the compact companion. Atmospheric perturbations, or surges in the flow, may initiate a major transfer of gas which is then accelerated to high velocity in the gravitational field of

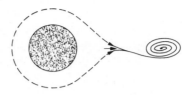

Fig. 8.11.   Illustration of gas flow into an accretion disk.

the compact star. As the gas streams toward the white dwarf it takes the form of a swirling whirlpool of gas known as an accretion disk as shown in Fig. 8.11. Frictional heating of the gas occurs in the disk, followed by a further high rise of temperature of gas that finally impinges on the surface of the white dwarf. Degenerate matter, because of its incompressibility, heats rapidly on impact to incite nuclear chain reactions setting the stage for explosive outbursts.

Novae, now known to be binary systems, provide an example of this process. A massive transfer of gas from a giant star onto a white dwarf initiates a nuclear explosion. The consequent bright stellar eruption is accompanied by the expulsion of an expanding nebular shell. An envelope of relatively dense ionised gas at a temperature around 10,000 K expands with an outward flow speed of $\sim$ 1000 km/sec. At first the optical luminosity increases as the envelope grows in size, and at maximum may attain 100,000 times the initial light output from the star. After a few days, as the emitting gas thins out the brightness diminishes and may take several years to resume its initial luminosity.

Two recent novae, Nova Delphini 1967 and Nova Serpentis 1970, afforded an excellent opportunity to study their radio emission. Regular observations commenced in 1970, so an almost complete radio history of Nova Serpentis has been recorded. It is clear that the radio output corresponds to thermal radio emission from the hot ionised gas. Compared with optical, for radio the expansion of the gaseous envelope must continue much longer before the density has fallen below that of a full radiator. Consequently over a year elapsed before the radio emission of Nova Serpentis reached its maximum value followed by a slow decline during the ensuing year.

We shall now pursue a discussion of the dramatic consequences of even more extraordinary states of condensed stars.

PULSARS

In radio astronomy we have become accustomed to surprising discoveries but none has seemed more astonishing than the pulsating radio sources first

detected in 1967 during a Cambridge research led by Hewish. An investigation was in progress of radio sources exhibiting scintillations imposed by the passage of radio waves through the irregularities of interplanetary ionised gas. The radio telescope consisted of a network of dipoles covering more than 10,000 m^2 of ground. It so happened that the large collecting area and long wavelength provided optimum conditions for the observation of pulsars. Nevertheless, it was the perception and perseverence of the research student, Miss Jocelyn Bell, that led to the totally unexpected discovery of these remarkable sources. When the occasional appearance on the records of pulses with a periodicity of about $1^1/_3$ sec was first noticed it seemed certain that man-made interference was responsible. Erratic variations in signal strength did not facilitate their elucidation, and they might easily have been dismissed as irrelevant to radio astronomy. When directional studies and continued analysis finally proved their celestial origin there remained the baffling problem: what type of astronomical source could possibly produce such rapid pulses and clocklike periodicity? Any suspicions that the pulses might arise from the machinations of extraterrestrial civilisations on distant planetary systems were ruled out, for other pulsars were soon discovered and their periodic Doppler shifts were found to be solely attributable to the motion of the Earth.

The first step was to decide what kind of astronomical object could emit such rapid and regular pulses. Despite variations of amplitude the precision of the repetition rate was remarkable. By 1968 the periodicity of the first discovered pulsar, labelled CP 1919 meaning Cambridge pulsar at right ascension 19 hr 19 min, was established at 1.3373011 sec with a stability of 1 part in 10^7. No normal star or even a white dwarf could vibrate or rotate so fast. Attention was soon focused on earlier theoretical concepts of possible further stages in the evolution of a collapsed star. In the 1930s Chandrasekhar had shown that if the mass exceeds 1.4 times that of the Sun the degenerate matter of a white dwarf could no longer be self-supporting. It was predicted that further contraction would ensue, the electrons and protons coalescing to form neutrons, hence producing a neutron star of perhaps only 10 km diameter. A spoonful of such material would have a mass of hundreds of millions of tonnes. To account for pulsars, Gold propounded the idea that they are rapidly rotating neutron stars emitting a narrow beam of radio waves, a concept that soon gained general acceptance. The high rate of spin is the expected result of the conservation of the angular momentum of the initial rotation of the star before its collapse. The contraction would also produce an intense magnetic field, possibly of the order of 10^8 T due to the compression of the original field of the star. The magnetic axis will probably be offset with respect to the spin axis, and it is assumed that a narrow beam is generated, sending flashes of radio waves to the observer in the manner of a lighthouse beacon. The process will be discussed later.

Before outlining the general properties of pulsars (over 300 have now been recorded) I must mention the exciting discoveries during 1968 of two very fast pulsars found within supernova remnants. One of these NP 0531 was detected in the Crab nebula by radio astronomers at NRAO, USA. With a period of 0.033 sec, measured at Arecibo, it was the fastest known pulsar. The period was also found to be lengthening by 36 nsec per day, an increase of 1 part in 2000 per year. The pulsar corresponds in position with a star that Baade and Minkowski in 1942 had deduced as the most likely residual star of the supernova explosion. In 1969 astronomers at the Steward Observatory, Arizona, began to search for optical pulses from this star. The outcome was the revelation that practically all the light is in the form of rapid flashes simultaneous with the radio pulses. How extraordinary that this faint star, known to astronomers for more than a hundred years, had not been shining continuously but had been rapidly flashing pulses of light. Subsequently, as techniques advanced, X-ray and γ-ray pulses were also detected. There is a striking similarity between the radio and the optical, X-ray and γ-ray average pulse profiles. Nonetheless, the discontinuity of the shape of the spectrum shown in Fig. 8.12 indicates that although closely associated, a different emission mechanism is involved. Finally, it was realised that whatever the precise process might be, the energy released by the slowing down of the rotation rate of the neutron star was more than sufficient to energise all the radiation from the Crab nebula. The problem of the continued replenishment of energy of the supernova remnant was now resolved.

It is interesting to recall that in the 1930s Baade and Zwicky had suggested that a neutron star would be the end-product of a supernova explosion. The conclusion that pulsars are neutron stars formed in

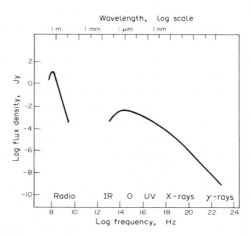

Fig. 8.12.   The spectrum of the Crab pulsar from radio to γ-rays.

supernovae was reinforced when a search for pulsars with the Molonglo radio telescope in Australia revealed a fast pulsar period 0.089 sec, at the centre of the Vela nebula, the brightest supernova remnant in the southern sky. The Vela pulsar was listed as PSR 0833-45 meaning pulsar at Right Ascension 08^h 33^m, Declination $-45°$, a designation now universally adopted: according to this system the Crab pulsar is PSR 0531 + 21. The Vela pulsar was observed to be slowing down at the rate of 10 nsec per day, equivalent to 1 part in 24,000 per year. The Vela remnant and pulsar are clearly older than the Crab. Optical search has revealed a faint blue star with part of the radiation pulsed, and although no X-ray pulsar source has been found, pulsed γ-rays have been detected. An interesting sideline of the radio observations was the occasional occurrence of a sudden jump in pulse rate followed by a very gradual return to its original value. Such discontinuities (often called "glitches"—an American word meaning "malfunctions") are believed to be caused by "star-quakes" on the surface of the neutron star.

We must now embark on a general survey of the properties of pulsars. At first it had appeared that the Crab and Vela must be prototypes of pulsars and their relation to supernovae. But the dramatic picture they present, although highly significant, now seems in certain respects to be unique because in other pulsars it has proved far more difficult to find evidence either of radiation outside the radio band or of close association with supernova remnants.

Pulse periods of most pulsars range from 0.033 sec to 3.7 sec, with a median value of 0.65 sec. Pulse lengths normally occupy a few per cent of the period. With no exception, pulse rates monitored over a long time-span are slowing down. Some show occasional glitches, as well as small fluctuations—the so-called restless behaviour. Nonetheless, the stability is remarkable, and the average rate of change over several years so predictable that some pulsars approach the time-keeping of an atomic frequency standard.

The ages of the pulsars can be roughly estimated from the pulse rates. Thus if P is the present period and $\dot{P}$ the rate of increase, then the age is $P/\dot{P}$ if we assume a uniform rate of change. However, the deceleration is greater in the younger, faster pulsars. Taking this into account $P/2\dot{P}$ is more nearly the true age. For the Crab pulsar this gives an age of about 1240 years, and although about 30 per cent longer than the historical age, the agreement is satisfactory. For the Vela pulsar the age is approximately 10,000 years, whilst for most pulsars the age is typically a few million years.

The distances of pulsars can be estimated by the influence of the intervening ionised interstellar gas on the arrival time of pulses at the longer wavelengths. It is well known that in an ionised gas the speed of travel of a radio pulse is slowed down by the scattering of radiation from electrons in the path and hence on the distance. Making the assumption

that ~10 per cent of interstellar hydrogen is ionised the first discovered pulsar CP 1919 (PSR 1919 + 21) was calculated to be about 400 light years away. Estimated distances of detected pulsars are mostly between 200 and 7000 light years. As the distribution of pulsars is concentrated towards the galactic plane, confirmatory evidence on the distances can be derived from the 21 cm line absorption due to the intervening atomic hydrogen in the spiral arms.

In contrast with the regularity of pulse recurrence, the amplitudes of the pulses are very variable and spasmodic. Despite this, the pulse shape averaged over many pulses is a well-maintained constant characteristic of each pulsar. For a large proportion the profile is double-humped. The pulse length typically occupies 3 per cent of the period, corresponding to a 10° beamwidth in a rotating source. The pulses exhibit strong linear polarisation with a field direction changing uniformly throughout the pulse profile. A good example is the Vela pulsar shown in Fig. 8.13 where the linear polarisation is nearly 100 per cent.

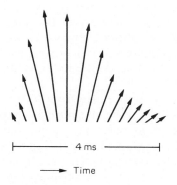

FIG. 8.13.   Changing polarisation during the Vela pulse.

Assuming the dimensions of the emitting region cannot exceed the travel time of a radio pulse across it, the radio brightness temperatures are of the order of 10^{24} K. Fine structure in the pulses indicates even higher temperatures, and in certain pulsars, such as the Crab, occasional giant pulses are emitted, and radio brightness exceeding 10^{30} K may be reached. The radio spectral index is typically ~ 1.5, and the linear polarisation tempts one to suggest that the synchrotron process is responsible, but the brightness temperature is impossibly high for normal synchrotron radiation. A coherent process must be involved to enable vast numbers of electrons to radiate together in phase or to amplify a generated wave. Incidentally, for the optical and X-ray pulses a coherent process is not demanded since effective temperatures are less than 10^{11} K.

Although various proposals have been put forward to explain how a rotating beam may be produced there is as yet no universally accepted model. Among important factors involved are the following. It is expected that the neutron star will be surrounded by ionised gas and that the powerful magnetic field forces the charged particles to co-rotate with the star. Owing to the high rate of rotation the particles at a certain radius must approach the speed of light. This radius defines a surface often called "the light cylinder". It is usually assumed that the magnetic dipole of the star will be inclined at an angle to the rotational axis. In several theoretical models it is considered that the emission originates where the magnetic field lines come close to the "light cylinder" and the relativistic speed directs the radiation forward in a narrow beam.

An inevitable consequence of the rotation of the neutron star is the production of the electromagnetic radiation beyond the "light cylinder" at the rotation frequency. The very strong fields at this comparatively low frequency would be efficient accelerators of charged particles. Much of the energy loss of the pulses probably occurs in this way and is believed to be responsible for initiating (in the Crab nebula for instance) the generation of almost all the radio, optical, and X-ray emission from the supernova remnant.

There is now the intriguing question, why are so few pulsars found in supernova remnants? The connection between the Crab pulsar and supernova is so firmly established, yet from the few hundred known pulsars and more than half as many supernova remnants the only undoubted associations are those in the Crab and Vela nebulae, and possibly in I C 443. Calculations based on estimated populations of pulsars and supernovae suggest comparable rates of formation in the Galaxy—about 1 per 30 years. We shall now examine reasons that may account for the paucity of observed coincidences.

Pulsars are mostly solitary objects, and a contributory factor could lie in the high velocities they can acquire during the supernova explosion. It is curious that the first intimation of the fast motion stemmed from a study of the variations of pulse amplitude. Many of the erratic fluctuations are undoubtedly intrinsic changes of emission at the source. Scheuer at Cambridge suggested that superimposed on the intrinsic variations there were scintillations on time scales of minutes to hours that could be due to the transmission of the radio waves through irregular ionised clouds in the interstellar gas. This indeed proved to be the case, and Galt and Lyne at Jodrell Bank, by recording at well-separated sites, demonstrated a fast-moving pattern of fluctuations at the Earth's surface implying a high transverse velocity in the pulsar. Subsequent interferometric studies of the proper motion of pulsars have confirmed the high speeds, typically near to 200 km/secs. By virtue of such velocities pulsars may become far displaced from their original positions and even beyond the confines of old remnants.

The disparity in ages also contributes to the lack of connection. For characteristic ages of pulsars are about a million years, whilst in comparison most supernova remnants have expanded and weakened sufficiently to merge into the interstellar medium in less than a hundred thousand years. Yet another factor is the beaming of pulsar emission which could mean that some pulsars remain unobservable if their beams are never directed toward the Earth. Nevertheless it seems odd that apart from the Crab none of the recent supernovae, Cassiopeia, Kepler, Tycho, can boast a pulsar. The absence of coincidence has raised the question as to whether the relation between supernovae and pulsars is as rigid as was first imagined. Might there be other types of stellar remnant of supernova explosions?

The role of binary systems in various forms of stellar activity has received much attention in the last decade. After all, roughly 50 per cent of stars are binaries. If one star undergoes a supernova explosion then it is probable that the component stars will fly off at high speed in opposite directions as a result of the orbital velocities and ensuing disruption. In very few cases would the system survive as a binary.

The first radio pulsar definitely established as a member of a binary system was the pulsar 1913 + 16, discovered at Arecibo in 1974. It had the second shortest period of any known pulsar, 0.059 sec. Although no radiating counterpart has been detected, the binary nature is revealed by cyclic Doppler shifts of the pulsar rate. The orbital period is 7.75 hours and the unseen companion is presumed to be a neutron star without observable radio emission. The mass of both components is estimated at about 1.4 times that of the Sun. This unique astronomical object has special significance in connection with tests to verify Einstein's theory of relativity as will be explained more fully later in Chapter 10.

X-ray Stars

X-ray observations from space vehicles have opened a new era in astronomical research. The first evidence of powerfully radiating X-ray stars marked the beginning of a succession of surprises. The discoveries have been particularly important in shedding new light on active phenomena in close binaries. We shall outline briefly the results of research on X-ray stars, and the part played by radio astronomy in establishing their properties.

The study of X-ray stars began in 1962 during a United States rocket experiment with the chance discovery of a powerful X-ray source in the constellation Scorpius. Another strong source in Taurus was identified as the Crab nebula during a lunar occultation in 1964. The launch of the first X-ray satellite in 1970 led to a new and richer phase in the detection of X-ray sources. The satellite was named Uhuru, meaning "freedom" in Swahili, in recognition of its launch from Kenya. The next X-ray satellite,

Ariel 5 launched in 1974, continued the survey of galactic and extragalactic sources. In recent years the Einstein Observatory satellite has brought a major advance in sensitivity and resolution.

SCORPIUS X-1

The first discovered X-ray star, Scorpius X-1, is the brightest object, apart from the Sun, in the X-ray sky. Four years after its discovery the position was determined accurately enough to establish optical identification with a peculiar faint blue star. The X-ray intensity shows erratic variations up to 50 per cent, and the optical brightness changes by as much as 100 per cent. Spectroscopic measurements have indicated that it is a binary star system with a period of about 0.8 day and is believed to comprise a normal type star with a neutron star companion. The X-ray emission corresponds to a thermal source at a temperature up to 100 million deg K. It is evident that gas surging from the visible star onto its compact companion generates the heat required to emit X-rays.

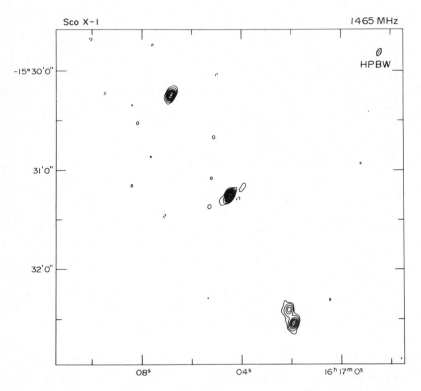

FIG. 8.14.   VLA radio map of Sco X–1 at λ = 20 cm. (After Geldzahler, Formalont, Hilldrup and Corey, 1981.)

Sco X-1 is associated with a remarkable radio source with a highly variable component centred on the star with two steady components situated in opposite directions 1'.2 away as shown in Fig. 8.14. All are non-thermal, but whereas the central component has a flat spectrum, the outer regions have steep spectra. These characteristics exhibit a striking resemblance to certain radio galaxies and quasars discussed later in Chapter 9. Undoubtedly Sco X-1 offers much scope for further research.

X-RAY PULSATING STARS

Certain X-ray stars have been found to radiate short-period regular pulsations of X-ray intensity at intervals of the order of seconds. Such behaviour is reminiscent of radio pulsars. These X-ray "pulsars" are not radio sources, but we must examine the properties of these unusual stars and explore their relationship to radio pulsars. Like other powerful X-ray stars they belong to binary systems. One of the sources, Centaurus X-3, has a pulsation period of 4.8 sec. Regular dips in output every 2 days indicate that it is an eclipsing binary. The source has been optically identified with a blue supergiant, and its unseen companion is assumed to be a neutron star. The X-ray emission will result from very high temperatures ($\sim 10^8$ K) produced in the accreting gas drawn to the neutron star. To explain the pulsations it is believed that the neutron star's magnetic axis is offset at an angle to the rotation axis. The accreting gas is channelled into the polar regions by the strong magnetic field of the spinning neutron star. The rate of spin therefore controls the pulsation period. Another similar pulsating X-ray star is Hercules X-1 with a pulsation period of 1.24 seconds, and an eclipsing binary period of 1.7 days.

Neither Cen X-3 nor Her X-1 are radio pulsars and it seems that the accretion regions are too dense to allow the escape of the radio waves. What X-ray and radio pulsars have in common is that both have pulsation rates controlled by spinning neutron stars. A particularly interesting difference is that in contrast with radio pulsars the X-ray pulsation rates are gradually speeding up. The X-ray behaviour is explicable as an increase of rotational energy of the neutron star would result from the inflow of accreting material.

The X-ray luminosities of the binary X-ray sources such as Cen X-1, Her X-1 and Sco X-1, are all very high, around 10^{30} W, about 1000 times the total radiated power of the Sun.

CYGNUS X-1

The most sensational deduction from any X-ray source followed the discovery and identification of the extremely variable source Cygnus X-1, which can show major changes in X-ray flux in less than 0.001 sec. A variable radio source was found to be associated with Cyg X-1 enabling the

position to be located accurately by radio interferometry. In this way the source was identified with a BO supergiant with an invisible partner. Assuming the supergiant to be roughly 20 solar masses, calculations indicated that the unseen companion must be nearly 10 solar masses. Now it has been shown that there is a mass limit estimated at about 3 solar masses above which a neutron star cannot support itself. A further stage of stellar collapse must then ensue so producing a "black hole", which means that gravitational attraction around the star has become so strong that no light can escape from it. The invisible partner of Cyg X-1 matched the required conditions; hence the excitement of this first convincing astronomical evidence of the existence of a black hole. The vicinity of the black hole must be the site of tremendous turbulent activity, a cosmic whirlpool, as gaseous material is sucked in at enormous speed from the surrounding accretion disk. Here the viscosity raises the temperature to 10^7 or 10^8 K, sufficient to emit X-rays as it spirals into the black hole. Surges and instability in the gas flow account for the rapid bursts of radiation.

CYGNUS X-3

The X-ray star Cygnus X-3 has the distinction of exhibiting the most spectacular radio flares. In a radio survey in June 1972, Cyg X-3 was reported as a weak source of variable radio emission, but in September of that year an enormous radio outburst increased the intensity by more than a thousand times. Curiously, there was no corresponding X-ray outburst. Since then, other major radio flares have been recorded, the source in between times lapsing back into extended periods of quiescence. During a flare the evolving spectrum of radio emission indicates synchrotron radiation from an expanding cloud of relativistic electrons, a conclusion confirmed by interferometric measurements. Unfortunately obscuration prevents optical identification. The source has been detected at infrared wavelengths, and both X-ray and infrared emission show a 4.8 day modulation indicative of a binary system.

Obviously radio stars and X-ray stars offer great scope for the study of extraordinary astronomical phenomena. We shall now conclude the discussion with a description of one of the strangest of stellar objects.

SS 433

During a survey of stars with strong emission line spectra, the American astronomers Stephenson and Sanduleak included in their list a star which they labelled SS 433. Meanwhile Clark and Murdin had started to examine the same star with the Anglo–Australian telescope. Their studies had been prompted by the coincidence of the star with a variable radio and X-ray source near the centre of the supernova remnant W50, for it seemed

possible that the star might have been the origin of the supernova. Clark and Murdin noticed that in addition to normal spectral lines there were strange emission lines they could not identify. The next stage proved particularly dramatic when a Californian astronomer, Margon, recognised in the spectrum of SS 433 bright emission lines of hydrogen and helium with huge Doppler shifts toward the red and the blue. The shifts indicated velocities more than 14 per cent of the velocity of light, both approaching and receding! Moreover, it was then realised that the Doppler shifts were gradually changing. As more data accumulated it transpired that the shifts were undergoing a regular variation. Red and blue shifts reached maximum values, then declined and reversed, the whole cycle repeating with a period of 164 days. Nothing similar had ever been seen before in emission line spectra. Strangely, the average of the shifts is centred not on zero but on a mean redshift corresponding to 4 per cent of the velocity of light.

At the same time, normal emission lines were also present in the spectrum. These were almost stationary, but when it was noticed that the wavelengths were subject to small 13 day periodic variations it became clear that the system was a binary, and these lines belonged to a comparatively normal star. The crucial problem was to explain the amazing Doppler shifts emanating from its companion star.

A model proposed by the Cambridge theorists Fabian and Rees fits the principal features of the strange spectra. According to this model two jets of glowing gas are ejected in opposite directions at speeds of about a quarter the velocity of light. We now have a picture, illustrated in Fig. 8.15, of the extended atmosphere of a star being sucked into an accretion disk around a compact companion, presumably a neutron star, where by some process not yet understood the gas is then shot out at huge speed in opposite directions. The Doppler shifts of the emission lines are due to the velocity components of the approaching and receding jets. The 164 day

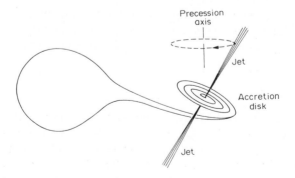

Fig. 8.15.   Model to illustrate accretion disk and jets in SS 433.

cyclic variation is attributed to a slow precession of the jets. In the crossover position both jets are at right angles to the line of sight. That the Doppler shift is not then zero is explained by the theory of relativity. At relativistic speeds, that is when they are an appreciable fraction of the speed of light, transverse motion is subject to an apparent time dilation causing a residual redshift. The twin-jet model has a notable resemblance to the jets found in radio galaxies and quasars discussed later in Chapter 9.

Radio astronomers have recently been endeavouring to map the structure of SS 433. The radio source is variable and subject to radio flares. Mapping with the aid of VLA and VLBI shows the existence of highly polarised extensions in structure correlated with the deduced positions of the optical jets.

The properties of SS 433 and its central position in the supernova remnant W 50 indicates that almost certainly in this case the supernova explosion has left not a pulsar but a quite different, peculiar stellar remnant. This throws into new perspective the questions on possible types of residual star and opens fascinating fields for further investigation.

Molecules in the Galaxy

For more than a decade following its detection in 1951, the 21 cm line of neutral atomic hydrogen was the only observed spectral line in radio astronomy. It had long been realised that other radio transitions would exist associated with molecular or ionised states of interstellar gas, but it was believed that the lines would be very weak because the abundance of the atoms or molecules would be small in comparison with atomic hydrogen.

Nevertheless it was realised that the radio studies of molecular lines could be a valuable guide to the chemical constitution and evolution of the dense clouds in the Galaxy which subsequently condense into stars and planets. Optical methods have been only a meagre help in this connection because of the scarcity of well defined visible molecular lines, and in any case obscuration is severe in gaseous nebulae. In contrast, energy changes in molecular rotation can activate a convenient range of radiant transitions at radio wavelengths. In consequence, many molecules are able to produce clearcut radio spectral lines. Moreover, very low energies suffice to excite many of the lines; for instance, temperatures from a few degrees to 100 K are capable of inducing radio transitions.

The detection in 1963 in USA of 18 cm wavelength lines due to hydroxyl (OH) molecules[1] reawakened interest in the astronomical potentialities of radio spectroscopy. OH lines were first detected in absorption looking towards the radio source Cassiopeia A, and the velocities deduced from

[1] There are actually four OH lines at 1612, 1665, 1667 and 1720 MHz.

the Doppler shifts were similar to those of atomic hydrogen in the same direction. Elsewhere in the Galaxy the pattern of similarity was not so well maintained, for the OH clouds responsible for absorption were found mainly in patchy, disturbed regions.

In 1965 the detection of OH lines in emission brought the most exciting results. The emission lines were first detected in the vicinity of ionised hydrogen (H II regions) surrounding very hot stars. The most remarkable features of these lines are their strong intensity, narrow bandwidth, and predominantly circular polarisation. In addition, studies with very long baseline interferometers revealed the sources to be extremely small, with angular sizes typically about $0''.005$, and radio brightness temperatures attaining 10^{12} to 10^{14} deg K. The linear dimensions calculated from the angular sizes are as small as about 10 to 30 AU, similar to that of the solar system.

Such extraordinary emission of intense, narrow bandwidth radiation from small volumes of gas indicates the occurrence of a special process of amplification comparable with a maser. In a laboratory maser the population of atoms at higher excited levels is artificially increased by a "pumping" process. In these conditions an incident radio emission of appropriate frequency can then trigger transitions from the overpopulated state to a lower level so releasing amplified "stimulated"" radiation. In the case of OH clouds, the pumping of the OH molecules to the higher excited states could be produced in various ways as will be mentioned later.

The OH emission so far considered is known as Type I, because subsequently it became clear that there are other types recognisable by their different characteristics. Type I associated with H II regions has predominantly strong lines at 1665 and 1667 MHz. These are the regions of star formation where gas and dust condense to form new stars. OH emission of Type II, with enhanced intensity at 1612 MHz, is found near old giant infrared stars. Another type, strongest at 1720 MHz, occurs in supernova remnants. The different types are evidently related to the physical conditions, and the various possible "pumping" mechanisms, which include infrared radiation, collisions, or chemical reactions.

In 1968 an even more striking example of an astronomical maser was discovered generating intense but variable radio line emission at 1.35 cm wavelength from water vapour (H_2O) molecules. The intensity and the very small angular sizes derived by long baseline interferometers, in some cases no more than $0''.0003$, implied radio brightness temperatures up to 10^{15} deg K.

A much studied example of interstellar maser sources is found within a complex H II ionised region known as W3. (It was first listed by the astronomer Westerhout from whom the nomenclature is derived.) The ionised region, embedded in obscuring dust, was mapped by its continuum radio emission. In one zone seventy components of OH emission grouped

in about a dozen main condensations have been located within an area of roughly 2″ diameter. A merit of line emission is that Doppler shifts reveal velocities. It has been deduced that the OH components are moving inward at about 6 km/sec representing remnants of accreting gas in continuing collapse towards a newly formed hot star. 7″ away is a strong H_2O maser source. It has been inferred that an H_2O maser is an early sign of gas condensing into a protostar.

It is evident that in order to produce the observed emission from such confined sources the gas densities must be very high. It has been estimated that the OH maser sources must have densities of the order of 10^6 molecules per cm^3, and the H_2O sources around 10^9 molecules per cm^3.

The improvements in millimetre wave receivers and radio telescopes quickly led to a spate of molecular line observations. Many normal lines were detected, for example from ammonia (NH_3), carbon monoxide (CO), cyanogen (CN) as well as more complex organic molecules such as formaldehyde, methyl alcohol, formic acid and so on. The list mounts each year. Such a multitude of molecular lines was totally unexpected, for it had not been anticipated that complex molecules could have formed and survived exposure to radiation in interstellar space. No doubt, dust plays an important shielding role in helping to preserve these molecules, whilst the simpler stronger-bonded forms like CO are able to survive more extensively. It is startling to realise that so many molecules known to have an important role in biochemical reactions are found to be so prevalent in regions where stars, and associated planetary systems (it may be assumed), are in the process of formation.

Carbon monoxide in the Galaxy has assumed special significance since line emission at a wavelength of 2.6 mm was first detected in 1970. The CO molecule can be regarded as a dumb-bell connecting carbon and oxygen nuclei. The 2.6 mm line is emitted when the molecule at its lowest energy level stops rotating. Excitation of the molecule into its rotational state is induced by collisions with hydrogen molecules which are some 10,000 times more abundant. Compared with hydrogen, carbon monoxide is the next most abundant species. As there is no radio line of molecular hydrogen (H_2), the widespread presence of CO is particularly valuable as an indicator of the principal distribution of H_2 and the cold molecular gas clouds which are the seat of star formation. Most of the CO is concentrated in clouds 10 to 100 light years in size, and contained within the galactic disk between 10 and 30 thousand light years from the centre. The Sun lies at the outer edge of this zone.

The profusion of molecular data is opening up a new vista in astrophysics and astrochemistry where the problem is to fit together a complicated multitude of clues into a unified concept of the behaviour of dense interstellar clouds and the essential processes of star formation. The amazing sequence of the synthesis of elements in stars, the subsequent

expulsion of stellar matter into space, the compounding of atoms into molecules in the cooler regions, and the condensation into new stars, is a subject of fascinating interest to which radio astronomy provides an important contribution.

Discovery of a Millisecond Pulsar

An exciting recent discovery is a pulsar with an amazingly rapid pulse rate. It had previously been thought unlikely that any faster than the Crab pulsar would be revealed. Not until techniques designed to detect very fast pulse rates were applied at Arecibo was it realised that PSR 1917 + 21 pulses at 642 times per second, a recurrence period of ~1.5 milliseconds. This implies a superfast spinning neutron star, and hence enormous rotational energy. The pulsation rate is remarkably stable, and we infer that the pulsar possesses an unusually weak magnetic field. Probably it is a very old pulsar for there is now no clear clue to its origin. No surrounding supernova remnant is apparent, nor evidence of binary motion; either may have been present long ago. One suggestion is that the pulsar was formerly in a binary system and it has acquired a very high speed of rotation from transference of energy by accretion from a companion that has now disrupted. Whatever its origin, PSR 1917 + 21 may well prove to be the first discovered example of an entirely new class of pulsars.

9. Radio Galaxies and Quasars

Classification of Normal Galaxies

We now come to the most challenging of problems, the nature of the universe beyond our Galaxy. How many galaxies are there like our own, and what other types of galaxies or astrophysical phenomena exist in these outer realms of space? Can we find other galaxies at various stages of development and so deduce how the universe has evolved? Out to what limits can we make observations at all? We shall be dealing with such questions in these last two chapters, and we shall find that radio is profoundly influencing our basic knowledge of the universe. Let us begin by looking at the optical picture and compare this with radio observations, first for galaxies and then for the remarkable objects known as quasars.

In the 18th century Sir William Herschel and his son John started a detailed telescopic survey of the huge system of stars in the Galaxy. In addition to plotting the stars, their observations added thousands of nebulae to those previously catalogued in Messier's list. Some of them were found to be the glowing gas clouds in the Galaxy discussed in the previous chapter. Most of the nebulae, however, were found to emit a continuous spectrum of light similar to the stars, and they were called white nebulae. A few could be accounted for as dust clouds in the Galaxy reflecting light from nearby stars. But the majority of the white nebulae could not be explained in this way, and it was surmised that they must be star systems or galaxies outside our own Galaxy, although too distant for the individual stars to be resolved. More support for this idea was obtained when Lord Rosse in the mid 19th century constructed his 180 cm diameter telescope. With its aid he obtained a clearer view revealing that many white nebulae have the shape of flat disks containing a spiral structure like that of Andromeda. The appearance of the Milky Way as a bright band of stars had previously suggested that our Galaxy must be in the form of a disk. Final confirmation that these nebulae are external galaxies, or "island universes" as they were then called, had to await the construction of more powerful telescopes with higher resolving power. The 2.5 m telescope at Mt. Wilson, USA, was completed in 1917, and the photographs obtained in the early 1920s by Hubble verified that Andromeda and other similar nebulae indeed contain myriads of stars. By recognising certain types of stars, such as the Cepheid variables, and measuring their apparent brightness, the distance to the nebulae could be estimated. In this way it

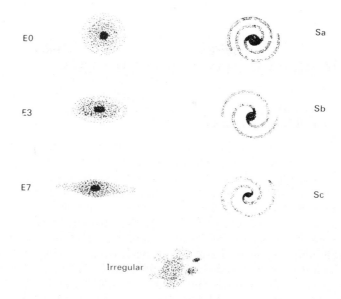

FIG. 9.1.   Diagram illustrating some types of galaxies in Hubble's classification. Ellipticals (E), Spirals (S), and Irregular.

was proved beyond doubt that most of the white nebulae are external galaxies viewed at great distances. Andromeda, the closest galaxy to us with a structure like our own, is at a distance of about 2 million light years. But this distance is small compared with many of the galaxies that can be photographed with large telescopes. The number of galaxies observable with the 2.5 m Mt. Wilson is so great that it is impracticable to catalogue them all. With the 5 m telescope at Palomar, USA, completed in 1948, more than a thousand million galaxies could be photographed extending to distances of about a thousand million light years.

A large number of galaxies that appear relatively bright in the telescopic surveys have been catalogued, and Hubble classed about 80 per cent as spiral, 17 per cent as elliptical, and 2 per cent irregular. Hubble's system of classification is illustrated in Fig. 9.1.

Our own Galaxy and Andromeda are both spirals of type Sb. Elliptical galaxies have an amorphous appearance with no spiral arms. The brightest and most massive ellipticals, class E 0, appear like huge spheres containing some million million stars. The series from E 1 to E 7 represents more flattened elliptical systems. Practically all galaxies possess a very bright core or nucleus, a central region with a very strong concentration of stars. In the E 0 galaxies, the nucleus is particularly small and bright. In addition

to elliptical and spiral classes, there are the less numerous irregular galaxies. Our nearest neighbour is of this type, the Magellanic Clouds, visible only in the southern hemisphere and named after the explorer Magellan, who observed them during his navigation of the southern seas.

It seems natural to associate spiral galaxies with rotation, for their disk-like shapes and spiral arms suggest massive spinning objects. As described previously in the chapter on the Galaxy, the spiral arms contain relatively high densities of gas and dust and young hot stars (called Population I). Older stars (Population II) are predominant in the galactic halo. Elliptical galaxies consist mainly of older stars (Population II) and usually not much gas or dust. It is generally believed that galaxies originated in the denser regions of primeval gas that pervaded cosmic space. As the galaxies contracted under their own weight concentrations of gas began to condense into stars. In elliptical galaxies most of the gas has evidently been used up in the process of star formation. In spiral galaxies it appears that rotation has counteracted gravitational contraction, and the swirling gas has been drawn into a disk and spiral arms. The older stars such as those in the galactic halo must have been produced early in the life of the galaxy before it settled down into its spiral shape. Formation of younger stars (Population I) then continues in the spiral arms.

The size and luminosity of galaxies span a wide range. The diameters of spiral galaxies mostly lie between 20 and 150 thousand light years, and their masses range from 10^9 to 10^{11} times the solar mass. The Sun is a convenient unit for reckoning the mass and radiation from galaxies. Our Galaxy is a large spiral with a mass equal to about 10^{11} suns and a light output equivalent to about 10^{10} suns. Andromeda is even larger and has about twice the mass of our Galaxy.

The elliptical galaxies show great diversity in size and luminosity. They range in size from dwarf galaxies about 2 thousand light years across up to giant galaxies 500 thousand light years or more in diameter. The most numerous galaxies distributed throughout space are those of low luminosity, the majority being small ellipticals. On the other hand, the giant ellipticals can be more luminous than any known spiral. We shall see later that further classification is necessary to describe some of the E type galaxies, particularly those with exceptionally bright or double nuclei.

If we examine the spatial distribution of galaxies we find that they are closer together than we might imagine when we think of the distance between stars compared with their size. For example, if we represent the size of a galaxy by a length of one metre, then on average the next galaxy would be about 100 m away. Galaxies are not uniformly distributed in space but tend to be grouped in clusters. The numbers in the clusters range from rich aggregates of 10,000 or more members, down to small clusters like the Local Group, which includes our Galaxy, the Magellanic Clouds, and Andromeda, and comprises about 30 members. There are two main

types of cluster. One kind has nearly spherical symmetry with greater concentration at the centre and is known as a regular cluster. The brightest members are nearly all large elliptical galaxies. The Coma cluster is a well-known regular cluster with over 800 members covering a circle of about 5° angular diameter in the sky. The other kind of cluster is the irregular type comprising many spiral as well as elliptical galaxies. Although the formation tends to be irregular, the elliptical members have a fairly spherical distribution concentrated towards the centre. The Virgo cluster is an example of an irregular cluster with over 1000 members contained in a 10° circle.

We have described galaxies in some detail because the question as to how galaxies form and evolve and how they are distributed throughout space is one of the most fundamental problems of astronomy. If we can solve this we have a key to the way the universe has developed. As we shall see later, radio as well as optical studies of certain unusual galaxies reveal remarkable phenomena of tremendous energy and may provide vital clues to the processes of formation and evolution of galaxies. However, let us first see how much radio and optical power is radiated by normal galaxies.

Radio Emission from Normal Galaxies

We may regard a normal galaxy as one fitting the standard optical patterns of Hubble's classification, and without peculiarities of structure or exceptional brightness. Radio observations of galaxies show that we can add another criterion. Normal galaxies have characteristic radio emission according to their class and luminosity. The Sb spiral galaxies like our own all have a similar radio output in comparison with their optical radiation. The same is true of other classes of spiral, although Sa galaxies are rather weaker radio emitters. Normal ellipticals (that is, excluding giants or those with unusual features) are very poor radiators at radio frequencies and few have been detected at all.

We can determine what radio and optical power is emitted by galaxies in the following way. If we find the radio spectrum of the source by observing at several frequencies, we can estimate the received power flux over the whole radio frequency band. Provided we know the distance of the source, we can then calculate the total power it is radiating. A similar procedure could be used at visible wavelengths, but here we can take in the whole optical band at once by receiving on an absorbing surface. The results we obtain then come out as follows.

The optical power radiated by the Sun is about 4×10^{26} W In comparison, the radio power is negligible; it is only about 10^{12} W (and therefore 100 million million times weaker than the optical power).

The light emitted by a spiral galaxy like our own is equivalent to about 10^{10} suns giving a total optical radiation of about 10^{37} W. The radio power,

however, has been stepped up enormously compared with that of 10^{10} suns, for the radio output of the galaxy is about 10^{31} W. Although it is still a million times less than the light originating from the stars of the galaxy, it represents an extraordinary amount of naturally produced radio emission. The radio band of frequencies is only a ten thousandth of the visible band, so the radio intensity expressed as power emitted per unit bandwidth seems even more remarkable. As shown previously (Chapter 2) this surprising amount of radio emission can be explained by synchrotron radiation from fast moving electrons in weak magnetic fields in the galaxy.

Before we describe radio properties of normal galaxies in detail it is of interest to note how much energy appears in the form of X-rays. The Sun radiates only about a millionth part of its power as X-rays. Certain stars, especially binaries, of types indicated in Chapter 8, are vastly more powerful X-ray sources. X-radiation from normal galaxies, amounting to about 10^{32} W, represents integrated emission from stars. Nevertheless, radiated power is dominated by the output of light from stars, exceeding the X-ray power by 100,000 times. In the radio band, stars contribute only an insignificant fraction compared with the synchrotron radiation arising from relativistic electrons in the interstellar medium.

The best known and most easily observed spiral galaxy outside our own is the Andromeda nebula, M 31, shown in Fig. 7.1. At the distance of 2.2 million light years, the tilted disk of the spiral subtends about 3½° by 1°. The radio spectral index of Andromeda is about 0.6 as compared with 0.5 for our Galaxy, and it seems clear that the radio emission must originate in a similar way by synchrotron radiation. Figure 9.3 shows a detailed radio map in the vicinity of the visible nebula obtained with the Cambridge One-Mile system with a resolution better than 2′.

Early radio maps of Andromeda had strongly suggested the presence of a radio halo extending far beyond the visible galaxy. More detailed recent examinations, however, have indicated that much of the surrounding emission is probably produced by unrelated radio sources and that a connection with Andromeda may be illusory. In contrast, an example of a spiral galaxy possessing a radio halo (similar to that deduced for our Galaxy) is illustrated in Fig. 9.2 showing the edge-on spiral galaxy NGC 4631 and the associated contours of radio emission mapped with the Westerbork radio telescope.

In 1850 Lord Rosse with his remarkable telescope first distinguished the spiral structure of an external galaxy, M51, the Whirlpool Nebula. The galaxy has the observational advantage of being seen face-on, although at 13 million light years distance it subtends only 10′. The Westerbork synthesis radio telescope possesses sufficient resolution to provide a detailed radio map of the spiral structure as shown in Fig. 9.4. The radio arms tend to follow the inside edges of the bright optical arms where there is a dust lane and compression of the gas and the magnetic fields.

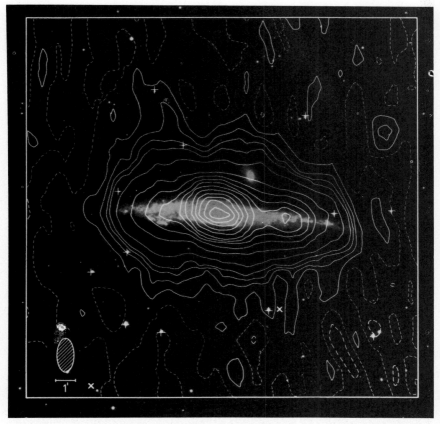

FIG. 9.2.   Radio halo at $\lambda = 49$ cm superimposed on photograph of the edge-on spiral galaxy
NGC 4631. (After Ekers and Sancisi, 1977.)

Synchrotron radiation is confirmed both by the spectrum and evidence of
linear polarisation.

M51 (NGC 5194) has a companion galaxy NGC 5195 shown in the upper
part of Fig. 9.4, and one can see how the radio map links the two galaxies
where there is scarcely any optical counterpart. It is also interesting to note
that there is no sign of a galactic halo. For both galaxies the region near the
nucleus is a concentrated source of radio and optical radiation.

The pursuit of such detailed studies of normal galaxies will undoubtedly
clarify the radio structure of the synchrotron emitting regions of fast
electrons and magnetic fields and their relation to the distribution of stars,
dust and gas. Are the supernovae remnants the source of fast particles?
Possibly so, but the radio structure certainly seems to extend beyond the
distribution of supernovae, which occur at a rate of about one per 30 years.

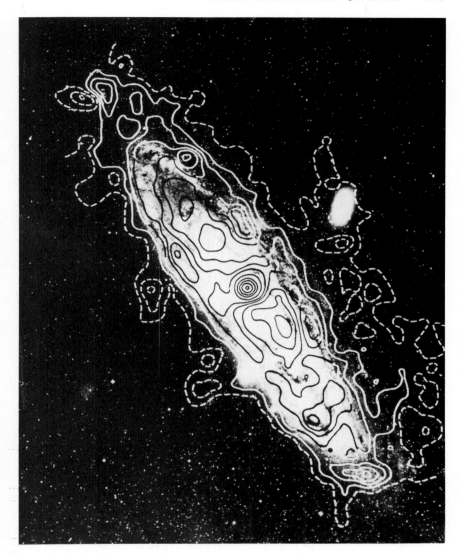

FIG. 9.3.   Radio map of the Andromeda Nebula at λ = 73 cm obtained with the Cambridge
one-mile radio telescope and superimposed on an optical photograph.
(After Pooley, 1969.)

The spiral arm structure may be explained by the combined influence of
flow activated from the galactic nucleus, rotation, and density waves in the
galaxy. These questions are a fruitful subject for continuing research.

The 21 cm line emission of atomic hydrogen can also be detected in
nearby external galaxies and the main structural features can be deduced.

FIG. 9.4.   Radio map of M51 (NGC 5194) and NGC 5195 obtained at 1415 MHz (λ = 21 cm)
with the Westerbork synthesis radio telescope. The radio map is superimposed on a
photograph taken by Humason with the Palomar 5 m telescope.
(After Mathewson, van der Kruit and Brouw, 1972.)

The radio data fit closely a type list of galaxies by de Vaucouleurs who
devised a detailed gradation of Hubble's classification. The total mass of
the neutral atomic hydrogen found in galaxies is also interesting. In
Andromeda it comprises rather less than 1 per cent of the mass of the
galaxy, and this is a typical proportion in spiral galaxies. The amount in

normal ellipticals is negligible. Only in the irregular galaxies is the proportion of neutral hydrogen very substantial, being of the order of 50 per cent of the total mass. It is therefore not surprising that irregular galaxies contain many young stars condensing out from the prevailing gas.

Radio Galaxies

We now consider more remarkable galaxies that are exceptionally intense sources of radio waves. The first discrete radio source to be discovered proved to be a very distant and unusual double galaxy, Cygnus A. Despite its great distance of 1050 million light years, the received flux density is stronger than from any other extragalactic radio source in the sky. This indicates at once that it is an extremely powerful radio emitter; in fact, its radio power output exceeds that of Andromeda by more than a million times. Cygnus A is actually emitting as much radio energy as it does light. Allowing for the wider visible bandwidth, we conclude that the radio intensity per unit bandwidth vastly exceeds that of the light.

Since the discovery of Cygnus A, many galaxies have been found with radio emission greatly exceeding that of normal galaxies. These very strong radio sources are called radio galaxies. Not all are as powerful as Cygnus A. We may define radio galaxies as those radiating from 10 to 10^6 times more than any normal galaxy. Only a very small proportion of galaxies are radio galaxies; but they are so powerful that their radio emission can be detected at great distances. Because many of them are so distant, first attempts to find optical objects corresponding to the radio sources met with little success. Only after the radio positions had been measured very accurately, and long exposure photographs with large telescopes had been taken, has it proved possible to identify many of the distant radio galaxies.

The identification of Cygnus A may be taken as a typical example. This source was first discovered during investigations of extra-terrestrial radio noise by Hey, Phillips and Parsons in 1946. With a beamwidth of about 12° they could only place the position of the source within about 2°. In the following years, more accurate positions were obtained using interferometers, but it was not possible to identify the source until 1951 when F.G. Smith at Cambridge obtained the first really accurate radio location at R. A. 19^h 57^m $45^s.3$ ± 1^s and Declination +40° 35′ ± 1′. Baade and Minkowski's search in this position with the 5 m Palomar telescope finally led to its identification with an unusual double galaxy, only faintly visible because of of its great distance. Originally thought to be a double galaxy, it is now surmised that a dust band gives its divided appearance.

Optical identification of radio sources is essential for any progress to be made in understanding them. In the first place, only by comparing the radio and optical properties can we realise the significance of the radio emission. Secondly, optical methods provide the only way of finding the

distance of a source, and it is imperative to know the distance to locate sources in space and to calculate how much power they are radiating.

We will here digress for a moment to consider briefly how the distances of galaxies are determined. Cepheid variable stars have been found to fluctuate with a period dependent on their luminosity. The Cepheid variable stars can therefore be used as distance indicators for galaxies that are near enough to distinguish individual stars. The intrinsic luminosity of Cepheid stars is known from their period, and the distances can then be deduced from their apparent magnitude. As Cepheid variables are not very bright stars, they can only be observed in nearby galaxies. Beyond this it has proved possible to estimate distances by making assumptions about the average absolute luminosity of the brightest stars in galaxies. Still larger distances are estimated with respect to the brightest galaxies in clusters. Distances can be inferred from the apparent magnitudes of these objects, because the fainter they appear the more distant they must be.

Using such criteria Hubble discovered a very remarkable law that spectral lines emitted by galaxies are shifted to longer wavelengths, by an amount proportional to the distance of the galaxy. The effect is known as the "redshift" because all the visible spectral lines are displaced towards the red part of the spectrum. A given line may have a certain wavelength, λ, measured in the laboratory. The same line from a distant galaxy is found to have a wavelength $\lambda + \delta\lambda$. Then the redshift, z, is expressed by the fractional change

$$z = \frac{\delta\lambda}{\lambda}$$

Hubble found that distance is proportional to the redshift, z.

The relation is often expressed as

$$D = \frac{cz}{H}$$

where D is the distance,

c is the velocity of light, 300,000 km/sec and

H is Hubble's constant, between 50 and 100 km per sec per megaparsec according to present estimates. Adopting the value 50 gives

$D = 6000z$ megaparsecs (Mpc)

or $D = 20,000z$ million light years (approx.)

The simple law derived by Hubble is valid when z is appreciably less than 1, but is expressed in a modified form for higher values of z. We shall discuss the meaning of the redshifts more fully later. We normally attribute a spectral shift to the Doppler effect due to motion, and a shift to longer

wavelength implies that sources are moving away. Hence the usual interpretation of Hubble's law of redshifts is that the universe is expanding.

Optical Properties of Radio Galaxies

At the time of writing the number of radio galaxies that have been optically identified exceeds 1000, so that it is possible to reach some general conclusions about their optical characteristics. The radio galaxies do not fall readily into Hubble's classification of the more abundant normal galaxies. The strong radio emitters show unusual features, such as intense and sometimes double nuclei or jets of glowing gas emitted from nuclei, or extensive gaseous envelopes. The radio galaxies correspond more closely to recent additions in the classification of galaxies. The further types include:

Seyfert galaxy, spiral with exceptionally bright nucleus.
D galaxy, elliptical type with extensive surrounding envelope.
DE (or ED) galaxy, intermediate between D and E.
DB (or dumbbell) galaxy, like D but with a double nucleus.
N galaxy, with brilliant starlike nucleus.

We shall consider later the remarkable quasi-stellar objects, the quasars and BL Lac types, that can be detected out to the farthest distance in the universe.

Normal elliptical galaxies have little gas, but those that are powerful radio sources usually show strong optical spectra characteristic of very hot gas. There is often evidence of high velocities indicated by Doppler shifts and broadening of the spectral lines. It has also been found that the identified radio galaxies are associated with intrinsically bright optical galaxies.[1]

These features of radio galaxies suggest that the galaxies may be in exceptionally active states. We shall illustrate this point more clearly later when considering the structures and properties of individual radio galaxies.

Spectrum and Polarisation of Radio Galaxies

The intense radio emission from radio galaxies at once rules out a thermal origin of the radiation. Two other criteria help us to decide on the emission mechanism, namely, the spectrum and the polarisation.

Most sources have a spectrum that can be roughly represented by power

---

[1] In drawing general conclusions one must beware of selection effects. The radio sources most likely to be identified are those which are strong radiators both at optical and radio frequencies. Nevertheless, the identified galaxies lie in a fairly narrow range of optical luminosity (mean absolute photographic magnitude Mpg = -20.5 ± 0.8) and generally brighter than giant spirals like our Galaxy or Andromeda.

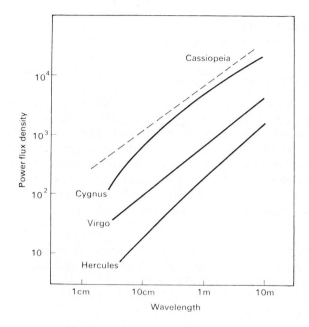

FIG. 9.5.   Spectra of radio galaxies Cygnus A, Virgo A, and Hercules A, compared with the supernova remnant Cassiopeia A. (Power flux density in units of 10^{-26} Wm^{-2}Hz^{-1}.) (After Dent and Haddock, 1966.)

flux P proportional to λ^x where each source has its own value of x called the spectral index. This relation plotted on a logarithmic scale is a straight line, and gives what is known as a "straight" spectrum. Some examples are shown in Fig. 9.5. Certain sources like Cygnus A have curved spectra because x is not constant over the spectrum. Figure 9.5 shows some examples of spectra of extended parts of radio galaxies. The compact regions in the vicinity of the nucleus will be considered separately.

The spectral indices of a large number of radio galaxies have been determined by measuring the power received at several wavelengths over the radio band. If we observe well away from the Milky Way, say at more than 10° away from the plane of the Galaxy, then we can be fairly sure we are looking at extragalactic sources. We can then plot the distribution of spectral indices of the large radio sources as shown in Fig. 9.6.

The median value of x is 0.7, and 50 per cent of the sources have spectral indices between 0.6 and 0.8. The similarity in the spectra of the sources indicates the same kind of radio process is responsible in all of them. It is interesting to note how well these spectra correspond with the non-thermal sources in the Galaxy like supernova remnants. The spectrum of Cassiopeia A is included in Fig. 9.5 for comparison. As explained in earlier

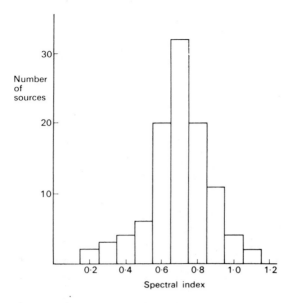

FIG. 9.6.   Distribution of spectral indices of radio galaxies. (After Conway, Kellermann and Long, 1963.)

chapters, the abnormal intensity and the spectrum can be accounted for by the synchrotron process, the radiation from very high velocity electrons in weak magnetic fields.

If the radio emission is synchrotron radiation we should expect it to be linearly polarised. As previously explained, depolarisation due to Faraday rotation and tangled magnetic fields can drastically reduce the percentage polarisation in the received flux especially at longer wavelengths. Until 1962, no linear polarisation had been observed in the radio emission from any extragalactic source. With improvements in technique even a small percentage of linearly polarised radiation can now be measured. Following the detection by Mayer and his colleagues in 1962 of about 8 per cent linear polarisation in overall emission from Cygnus A at 3 cm wavelength, it has been found that nearly all radio galaxies exhibit a small but significant degree of polarised emission. As may be expected, the percentage polarisation is greater at short wavelengths. With the angular resolution now available the polarisation of different parts of the source can be distinguished. The true polarisation from each region may then be deduced by making measurements at several wavelengths and extrapolating the graph to zero wavelength as illustrated in Fig. 9.7. In this manner the effect of rotation on the direction of polarisation during propagation can be eliminated.

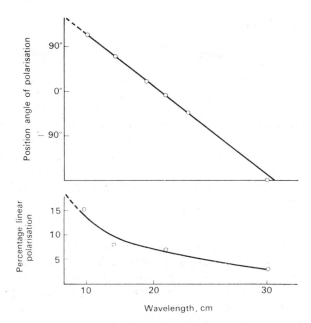

FIG. 9.7. The rotation and loss of polarisation increases with wavelength. The above measurements for a central region of Centaurus A show how the intrinsic polarisation, corresponding to zero wavelength, is deduced by graphing observations at various wavelenghts. (After Gardner and Whiteoak, 1963.)

All the evidence indicates that the main cause of the radio emission from galaxies is the synchrotron process, and the same is true of sources like supernova remnants within the Galaxy. The principal properties are common to all, namely:

(a) very high intensity of radio emission;
(b) spectra showing increase of intensity with wavelength;
(c) the emission shows linear polarisation.

The most remarkable feature of the strong radio galaxies is the vast scale and power involved. The radio galaxies are up to a million times more powerful radio transmitters than normal spiral galaxies. Within the spiral galaxies, the high energy electrons and magnetic fields required for synchrotron radiation might originate from supernovae. In the radio galaxies there is evidence of unique sources of energy. As we shall see from detailed studies of structures and changes in central regions it appears that exceptional activity in the dense nucleus is responsible for generating the high energy particles and fields producing the remarkably intense radio emission.

Structure of Radio Galaxies

Let us now consider the size and structure of the radio sources to see how they are related to the visible galaxies. The simplest way is to have a large radio telescope with a single, narrow beam to scan the source and so build up a picture of the radio distribution. Unfortunately the beamwidths of steerable radio telescopes are too wide in comparison with the angular size of most radio galaxies. A radio galaxy of exceptionally large angular size is Centaurus A in the southern hemisphere. This source subtends several degrees and there are two reasons why it appears so large: it is the nearest strong radio galaxy, and the radio source is also very much bigger than the visible galaxy. Most radio galaxies subtend minutes of arc or less, and the size and structure have to be deduced by making observations with interferometers or other aerial systems employing widely-spaced aerial elements.

We will now look at the contours showing the intensity distribution of radio emission from selected radio galaxies. The radio pictures illustrate well the fact that the radiation mostly originates from regions extending far beyond the visible galaxy. The fast electrons and magnetic field have evidently expanded out from the visible galaxy until they have been slowed down or stopped by the rarefied gas that pervades space outside the galaxies. There is of course no reason to expect that the fast particles would produce visible radiation. Only if they have extremely high energy could they produce optical synchrotron emission, and then only for a short time because the rate of loss of energy would be so rapid. In the few cases where there is evidence of light being emitted by this process there must be continual replenishment of the very fast electrons.

Cygnus A

Cygnus A, the first discovered radio source beyond our Galaxy, has long presented an intriguing test of observational techniques even though in received radio intensity it is the strongest extragalactic source. At a distance of 1050 million light years it is a most energetic radio galaxy radiating about 10^{38} watts of radio power. But owing to its great distance, its angular subtension is only of the order of a minute of arc, so that very fine resolution is required to map its radio structure. More than six years had to elapse after its discovery in 1946 before it was shown to be a double radio source associated with an unusual optical galaxy lying midway between two large regions of radio emission. The optical object is a giant D type galaxy straddled by a dust band, the brightest member of a rich cluster of galaxies. Aperture synthesis methods have made it possible to derive the radio distribution in some detail. Figure 9.8 shows the radio structure mapped at 6 cm wavelength by the Cambridge 5 km radio telescope with a

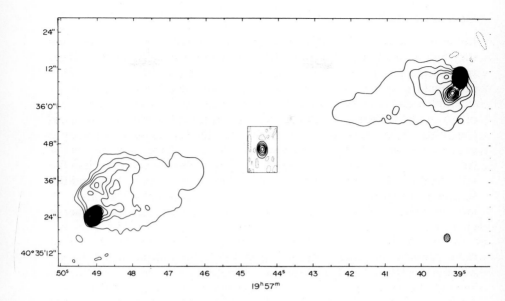

FIG. 9.8.   Cygnus A at λ = 6 cm mapped with the Cambridge 5 km radio telescope. The contour interval is 10,000 K for the two main components, and 2000 K for the weaker component centred on the optical galaxy. (After Hargrave and Ryle, 1974.)

resolution of 2″ by 3″. Special features are the intense heads or "hot spots" near the extremities of the outer lobes, and a comparatively weak but very compact source coinciding with the optical galaxy. Very long baseline interferometer (VLBI) measurements to explore the central compact source have revealed a core of about 10 light years diameter with an extension towards an outer lobe. The spectral index is near zero, in contrast with the steeper spectral indices of about 0.8 for the hot spots and 1.2 for the diffuse outer lobes. The hot spots and the compact core centre on the galaxy lie in the same line. This strongly suggests that the outer regions must be energised by beams of particles and fields ejected in opposite directions from the nucleus of the galaxy. The concentrated zones in the outer regions are then formed where the beams are contained by the pressure of the surrounding medium. A crucial question concerns the process by which synchrotron emission is generated and maintained at such vast distances from the nucleus, for the separation of the outer lobes is about 500,000 light years. Consequently the idea has gained ground that the nucleus is the "engine" continually or repeatedly feeding energy to the outer parts of the radio source. As we shall see, this concept is reinforced by examination of the structures and properties of other radio galaxies.

Centaurus A

Centaurus A, observable in the Southern hemisphere, is the nearest strong radio galaxy. At a distance of about 15 million light years, this huge radio source subtends 8° by 4° in the sky. Due to its large angular size it was the first radio galaxy for which the structure and polarisation could be examined in detail.

The optical galaxy associated with the radio source is a peculiar elliptical galaxy NGC 5128, about 50 thousand light years in diameter. Optically it is one of the brightest galaxies in the sky. The galaxy has a striking dark band of dust across it as shown in Fig. 9.9(a). The radio distribution consists of an extensive double source, shown in Fig. 9.9(b), at least 50 times larger than the optical galaxy, and over 2 million light years in extent.

The apparent magnetic field distribution plotted in Fig. 9.9(c) has been derived from the linear polarisation of the radio emission. The percentage polarisation at 10 cm wavelength is typically 10–20 per cent; in places it reaches as much as 40 per cent. Allowance for Faraday rotation has been made by measurements at different wavelengths, as shown in Fig. 9.7, in order to derive the direction of the polarisation of the electric field of the emitted radiation. According to the synchrotron theory, the magnetic field is at right angles to this polarisation. It is notable that the magnetic field

FIG. 9.9 (a) A photograph of the optical galaxy, NGC 5128.

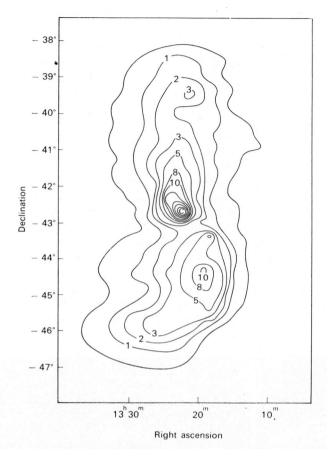

FIG. 9.9. (b) Radio map of Centaurus A at λ = 21 cm. The numbers on the contours are radio brightness temperature in deg K. The inner contour of the central source (C) is 100 K, and the optical galaxy lies in this region.

structure is quite uniform over large areas of the source, indicating fairly simple types of field structure.

The radio power emitted by Centaurus A is about 10^{35} W, and is therefore 1000 times less powerful than Cygnus A.

Approximately 20 per cent of the total radio emission comes from a central component comprising a small but intense double source conciding approximately in position with the optical galaxy as shown in Fig. 9.9(c).

The existence of inner and outer double radio sources suggests repeated events in the replenishment of the high-energy electrons and magnetic fields required to produce synchrotron emission. The diffuse outer lobes are not, however, aligned with the inner double source,

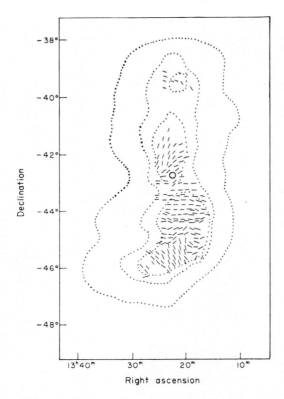

FIG. 9.9. (c) Magnetic field directions in Centaurus A deduced from radio observations. The outer shape of the radio galaxy is shown in dotted contours. The optical galaxy NGC 5128 is indicated by a circle. (After Cooper, Price and Cole, 1965.)

reflecting changes in the past history of the galaxy. In fact, peculiar optical features have given the impression that a massive merging of surrounding gaseous material may have occurred. The diffuse outer radio lobes are probably the fading remnants of previous very powerful radio emission. Closer to the nucleus, lingering activity is evident, possibly a consequence of continuing infall of gas into a massive but very compact nucleus. The region of the inner radio lobes is also a source of X-rays believed to originate from the inverse Compton action of relativistic electrons colliding with radio photons (see p. 23). In addition, in one direction from the nucleus a narrow X-ray jet leads toward an optical jet, both being in line with the inner radio lobes. The jets are thought to stem from very hot plasma (up to 10 million deg K) ejected from the nucleus; a corresponding radio jet has been discerned.

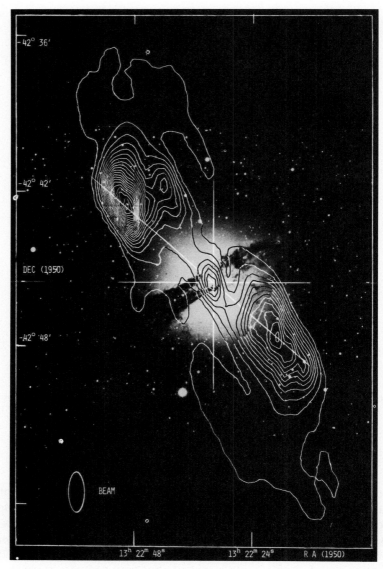

Fig. 9.9 (d) High-resolution map obtained with the Fleurs synthesis radio telescope at λ = 49 cm of the central region of Centaurus A in relation to the optical galaxy. (After Christiansen *et al.*, 1973.)

The nucleus reveals itself as an extremely compact source detectable at infrared, millimetre, X-ray and γ-ray wavelengths. Variability is indicative of the very small size of the source. VLBI measurements of the radio core give a diameter less than a millisecond of arc radiating strongly at short wavelengths. The infrared dimensions of the nucleus are no more than a few light months. It is clearly a centre of extraordinary activity.

Virgo A

Virgo A is the nearest radio galaxy in the northern sky. It has a radio structure of the type described as "core and halo" where a central intense region is embedded in a much wider distribution. The associated optical object is a very bright elliptical galaxy, M 87 (NGC 4486), a prominent central member of the well known Virgo cluster. Its most remarkable optical feature is a bright blue jet emerging from the nucleus as shown in Fig. 9.10(a). The distance of the galaxy is about 50 million light years; the extension of the visible jet is about 20″, corresponding to a length of about 5000 light years. In long exposure photographs, the optical halo of the galaxy, with a diameter of over 100,000 light years, masks the jet from view. The jet possesses a marked degree of linear polarisation, up to 30 per cent, providing strong evidence that the optical radiation from the jet is produced by the synchrotron process.

The radio intensity, polarisation, and spectrum of Virgo A show that the radio emission is synchrotron radiation. The structure comprises a core, consisting of the nucleus and jet, embedded in a radio halo covering 12′ by 15′, somewhat larger than the optical halo. Because their radio spectra are different, the core is best observed at short wavelengths, and the halo at longer wavelengths, the radiation from each being equal at about 40 cm wavelength. It is curious that the combined spectrum shown in Fig. 9.5 turns out to be such a straight line.

Recent Westerbork radio maps indicate that the radio halo really has a double structure, with one large lobe almost overlapping the other. This demonstrates an important point when we are considering the apparent shape and extent of radio galaxies, namely their dependence on how they are inclined to the line of sight. In most cases knowledge of the inclination is uncertain. The comparatively small proportion of core-halo structures could, in some cases at least, easily be double sources viewed near end-on.

The radio nucleus and jet closely coincide with their optical counterparts, although the jet is longer at radio wavelengths. VLBI observations of the nucleus at cm wavelengths show that the angular size is less than 0″.001, equivalent in linear dimensions to only a few light months. The intensity is surprisingly steady for such a concentrated source with an emergent jet. The radio nucleus and jet mapped at λ = 6 cm with the Very

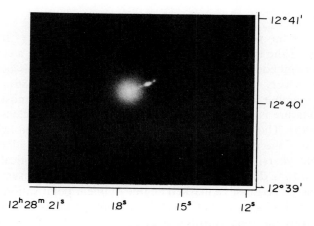

FIG. 9.10. (a) Photograph of the nucleus and jet of M 87 (NGC 4486) in the Virgo cluster.

Large Array (VLA) is shown in Fig. 9.10(b). The jet radiation at both radio and optical wavelengths arises mainly from several knots of emission. We assume that the jet is a beam of ionised gas ejected from the nucleus at very high speeds. Relativistic electrons in magnetic fields give rise to synchrotron emission. However, as the lifetime of electrons emitting optical synchrotron radiation is only ~ 20 years it is clear that a continual process of generating and replenishing relativistic electrons must occur at the radiation zones along the jet.

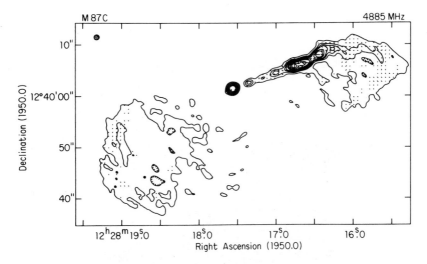

F<small>IG.</small> 9.10.(b)   VLA radio map at λ = 6 cm showing the core and jet of Virgo A. (After Owen and Hardee, 1980.)

M 87 was the first external galaxy known to emit X-rays. It is a large source covering almost 1° and probably due to thermal radiation from a gaseous halo at a temperature more than 10^7 K (an X-ray halo is similarly associated with Cygnus A). M 87 is the brightest galaxy in the Virgo cluster, and extensive X-ray emission indicating the presence of extremely hot gas has been found to be a notable characteristic of large clusters of galaxies. X-ray emission from the vicinity of the M 87 core has also been detected. The radio and optical power radiated by M 87 approaches 10^{36} W, although the luminosity associated with the galaxy is dominated by the X-ray output amounting to 5×10^{36} W.

Double Structures and Jets

The improvements in angular resolution in recent years have brought detailed radio structures into much clearer focus. We shall now show further maps of radio galaxies to illustrate characteristic types.

The source 3 C 111 provides an example of close similarity between inner and outer double lobes. The optical galaxy, with a bright core giving the galaxy an almost starlike appearance, has a redshift $z \approx 0.5$, corresponding to a distance of 10,000 million light years. Figure 9.11 shows the large-scale lobes mapped at λ = 21 cm at Westerbork, and the core structure at λ = 2.8 cm determined by VLBI. The core comprises two concentrated lobes separated by 0″.006 having radio brightness tempera-

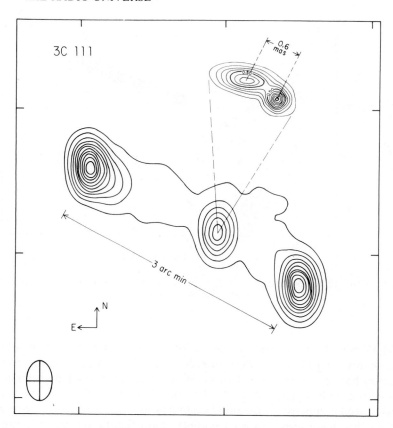

FIG. 9.11.   Radio maps of extended and compact components of 3 C 111. The larger structure was obtained at λ = 21 cm by Högbom and Carlsson, 1974, with the Westerbork array, and the central region at λ = 2.8 cm by Pauliny-Toth *et al.* using VLBI, 1976. (After Kellermann, 1978.)

tures, although variable, of ~ 10^{11} K. The alignment and similarity in shape between inner and outer lobes provides remarkable evidence of the same mechanism of formation and alignment being maintained over an interval of more than a million years.

A fine example of a different type of symmetrical double structure is shown by the map of radio source 3 C 449 in Fig. 9.12. The source has been identified with a DE galaxy having a redshift $z = 0.018$. The radio structure well illustrates the nucleus and the emerging double jets feeding diffuse outer lobes. The twists in direction may be attributed to precession of the axis of the nucleus, or possibly to gravitational encounters with nearby galaxies.

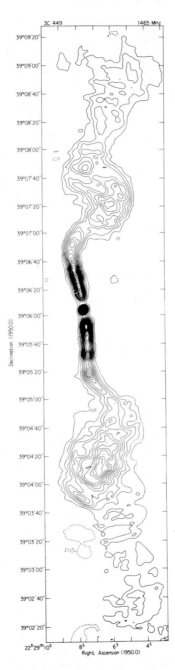

FIG. 9.12.   VLA map at λ = 20 cm of the radio galaxy 3 C 449. (After Perley, Willis and Scott, 1979.)

A spectacular radio jet in a huge radio source has been found associated with the elliptical galaxy NGC 6251, redhift $z = 0.023$, corresponding to a distance of about 500 million light years. The radio structure, first mapped by the Cambridge group, showed the large-scale radio source to be an irregular double covering a total extent of almost 10 million light years; but the most striking feature is the exceptionally long one-sided jet extending for almost half a million light years from a compact core centred on the

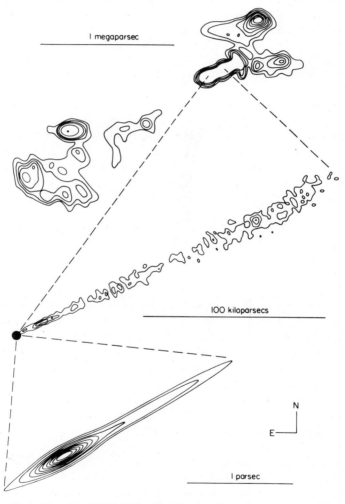

Fig. 9.13.   The radio galaxy NGC 6251. Uppermost is the large-scale structure at $\lambda = 21$ cm, and in the centre the huge jet emanating from the nucleus mapped at 11 cm by Waggett, Warner and Baldwin, 1977, with the Cambridge half-mile and 5 km radio telescopes. The nuclear component derived by VLBI at $\lambda = 2.8$ cm is shown below. (After Readhead, Cohen and Blandford, 1978.)

optical galaxy. Subsequently, VLBI observations of the core by the Cal. Tech. group showed an elongated structure looking almost like a cosmic blow-torch, in the direction of the long jet. The components of this remarkable radio source are illustrated in Fig. 9.13.

Double structures and jets aligned with the nucleus, so often found in radio galaxies, lead almost inevitably to the conclusion that the beams producing the radio emission must be ejected in opposite directions along the rotation axis of the nucleus. It appears certain that the jets are continually feeding energy to the extended structures. The occurrence of inner and outer lobes may be attributed to repeated explosive events. The nucleus, although extremely small, must supply enormous power. As will be discussed later, the release of such energy may be explained by gravitational collapse leading to the possible formation of a black hole at the nucleus.

We mentioned earlier how the alignment of structural features with respect to the line of sight may affect their appearance. This provides a clue as to why jets sometimes appear one-sided. Quite probably such jets may in fact be double ejections in opposite directions, but the relativistic beaming of the radiation enhances the approaching jet while diminishing the receding jet so as to render it practically unobservable.

We shall now consider other examples of radio galaxies firstly to demonstrate evidence of activity at the core, and secondly to show how structures may be altered when galaxies move at high speeds through the surrounding medium.

Perseus A

The optical galaxy NGC 1275 is the brightest member of the Perseus cluster, and an exceptionally active galaxy with a starlike nucleus. The galaxy has often been classified as a Seyfert, but morphologically it more closely resembles an elliptical. The redshift $z = 0.018$ indicates a distance of 360 million light years. The corresponding radio source, Perseus A (3 C 84), has a complex core surrounded by a halo 5' or more in extent. Within the core, a very small nucleus of varying intensity is one of the brightest centimetric sources in the sky. The power contributed by the three principal components can be discerned in the spectrum shown in Fig. 9.14. The total radio power is about 10^{35} W.

The most fascinating region is the nucleus. The steep fall-off in spectral intensity at wavelengths longer than a few cm is characteristic of a source sufficiently condensed to cause self-absorption of emitted radiation and indicates an angular size $\sim 0''.001$. Variability is a further pointer to very small size and remarkable activity. But the results of intercontinental very long baseline inferometry have revealed the most interesting information of structural changes. Examples of radio maps obtained in the years 1972–6

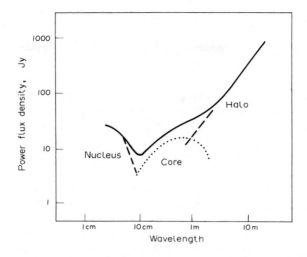

FIG. 9.14.   Spectrum of the radio galaxy Perseus A showing the total flux density and the three components. (After Ryle and Windram, 1968.)

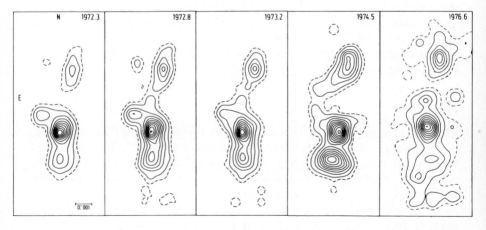

FIG. 9.15.   Structural changes in the nucleus of NGC 1275 (Perseus A) as shown by VLBI radio maps at $\lambda = 2.8$ cm. (After Preuss *et al.*, 1979.)

are shown in Fig. 9.15. The separation of the outer components is about 10 light years but a significant part of the emission from the central component originates from a volume no greater than a few light months. Both the rate of increase of power and the expansion of the components suggest an explosive event occurring some 20 years ago.

NGC 1265

Another radio source of special interest in the Perseus cluster is associated with the galaxy NGC 1265. A common characteristic of rich clusters is the rapid motion of constituent galaxies under the influence of gravitational forces. Toward the centre there is a more tightly packed distribution of galaxies, and relatively dense gas at a temperature of many million deg K has been recognised by its X-ray emission. In consequence, galaxies moving at high speeds, over 1000 km/sec, plough their way through dense intergalactic gas. For a fast-moving galaxy, the motion through the cluster medium modifies the source structure constricting the front by dynamic pressure and leaving a following trail of particles and fields. The radio galaxy of NGC 1265 provides an example of such a "head–tail" formation. Figure 9.16 shows a radio map of the source superimposed on a photograph of the optical galaxy. The symmetry is clearly reminiscent of double radio source structures except that the two parts are swept back along the trail. The polarisation mapped at Westerbork is shown in Fig. 9.17 and a striking uniformity is apparent in the magnetic field distribution. The analogy between the trailing configuration and the magnetospheric tail of the Earth in the solar wind is particularly notable.

Size and Energy of Radio Galaxies

The sizes of radio galaxies mostly range from about 100,000 to a million light years. At the time of writing the largest known radio galaxy is 3 C 236 with elongated double lobes covering a total extent of about 18 million light years, and having a compact radio core optically identified with an elliptical galaxy of redshift $z \approx 0.1$. Since the particles and fields transporting the energy to the outer components cannot travel faster than light they must have been ejected from the nucleus some 10 million years ago. Despite the vast scale and age, the outer extremities of the lobes retain radio-bright rims and remarkable alignment with the optical galaxy. The radio core, mapped with the aid of VLBI, is elongated in the direction of the outer lobes. The radio luminosity of 3 C 236 is about 2×10^{36} W.

Let us now consider the extraordinary energy associated with radio galaxies. There are two ways of estimating the energy. Firstly, we note that the separation of the two radio components may extend to the order of 10^5 to 10^6 light years. Even if they had been expelled at speeds close to that of light, they must have existed for times approaching a million years. As 1 year is roughly equal to 3×10^7 seconds, then if we assume that a very strong radio galaxy has been radiating 10^{38} W for a million years the total energy radiated is $10^6 \times 3 \times 10^7 \times 10^{38} = 3 \times 10^{51}$ joules. If we suppose that between 0.1 and 1 per cent of the total energy has been converted into radio waves, we find that the energy of the fast particles and fields must have exceeded 10^{54} joules.

FIG. 9.16. Radio map of NGC 1265 at λ = 6 cm obtained with the Westerbork synthesis radio telescope. (After Wellington, Miley and van der Laan, 1973.)

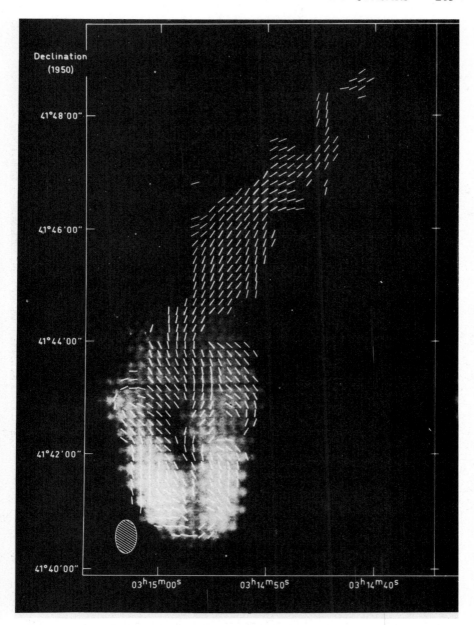

Fig. 9.17. The magnetic field distribution in the radio galaxy NGC 1265. (After Miley, Wellington and van der Laan, 1975.)

Alternatively, we can estimate the energy of radio sources as follows. Knowing the intensity of emission and the size of the source we can calculate the energy of the fast electrons and the strength of magnetic field required to produce the observed synchrotron radiation. We assume that the protons (hydrogen ions) have also been accelerated to high speeds like the electrons. Then we add up the total energy of the particles and magnetic fields and again arrive at a figure of about 10^{54} joules.

There is a famous law due to Einstein telling us that the energy possessed by matter is given by

$$E = mc^2$$

where m is the mass, and c the velocity of light. Nuclear reactions, such as those that occur in the Sun (and in atomic bombs) are ways of converting mass into radiant energy. Now the mass of a galaxy is equal to about 10^{11} suns. This gives a total mass energy $E = mc^2 = 10^{58}$ joules. Hence the radio emission shows that at least 1/10,000 of the total mass energy of the galaxy, equivalent to 10 million suns has been transformed into the energy of particles and fields giving rise to the radiation. No known nuclear process can produce this wholesale conversion of energy. The most likely source of energy is gravitational collapse, the central mass of the galaxy converging into a small volume with continually increasing speed under the force of gravity. It is then necessary to explain how the energy can be converted into that of particles and fields giving rise to the radio emission. We shall consider the transformation mechanism again later in the Chapter. Several galaxies are known to radiate as much as 10^{38} Watts of radio power, similar to Cygnus A. Although the majority of radio galaxies are somewhat less energetic, nonetheless they all manifest to some degree the same energy problem which becomes even more acute when we add the power radiated in other spectral bands such as X-rays. But first we must discuss the quasars which can radiate even more powerfully than radio galaxies.

The Discovery of Quasars

The discovery of quasars was one of the most exciting events in the history of radio astronomy. In 1960 the position of the radio source 3 C 48 was pinpointed with high precision (to about 5") by means of the radio interferometer at the California Institute of Technology, USA. Measurements at Jodrell Bank of the amplitude of interference fringes obtained with very long baseline interferometers had shown that this source must have a very small diameter, less than 1". Aided by the accurate radio position, Sandage took photographs of the region with the 5 m Palomar telescope and identified the source with what appeared to be a star having unusual spectral characteristics. The continuum light of the star was

exceptionally blue, and the spectral lines were puzzling as they were unlike those found in other stars. At that time it was believed that all discrete sources were extragalactic, and the observations were initially hailed as the first real evidence of a radio star in the Galaxy. The mystery of the spectral lines was not resolved until 1963 after observations of a similar source. In this case, the position of the radio source 3 C 273 was determined very accurately (to within 1″) during a lunar occultation, that is, when the Moon's path in the sky happened to occult the source from view, a technique of position finding developed by Hazard. 3 C 273 was found to consist of a compact source, less than 1″ in size, and an extended region about 20″ away. Hazard and his colleagues noticed that the compact source appeared to coincide with a 13th magnitude star. Again, observations with the 5 m Palomar telescope revealed a starlike object with mysterious spectral lines. Then came the dramatic discovery by Schmidt that the spectral lines had not been recognised because they had unexpectedly large redshifts. The value of z, the fractional increase in wavelength of the spectral lines of the starlike object associated with 3 C 273 was found to be 0.16 (that is, the wavelengths of all the lines were increased by 16 per cent). Re-examination of the spectrum of 3 C 48 showed an even larger redshift of $z = 0.37$. So these sources were not stars after all, but according to Hubble's law of redshifts enormously distant objects—at about 3000 and 7000 million light years respectively. It was concluded that as they could be observed at such great distances they must be extremely bright, with an optical luminosity over a hundred times that of any other type of galaxy. At the same time, their starlike appearance indicated they must be very small compared with other galaxies. In the following years many other radio sources have similarly been identified with intensely bright starlike objects with high redshifts, and they are known as quasi-stellar radio sources, or "quasars" for short.

Properties of Quasars

3 C 48 and 3 C 273 are interesting examples, possessing many typical properties of quasars. In both there is a very bright optical source radiating particularly strongly in the blue part of the spectrum and extending into the ultraviolet. The spectrum is often described as having an "ultraviolet excess". The light from both sources also shows remarkable variations of intensity. On looking through old photographic plates it was found that the starlike object associated with 3 C 273 can be discerned on sky photographs taken during the past 80 years, and although the average brightness remains unaltered, considerable short period variations have occurred. On some occasions changes in brightness by 2 or 3 times have occurred in periods as short as a month. The large fluctuations imply that the source must be extremely small because no disturbance can be propagated

throughout the source at a speed faster than light. We conclude that the starlike object cannot be more than a light year in diameter!

The spectra of both sources show emission lines indicating the presence of hot gas. The spectral lines are broad, and assuming the breadth to be caused by Doppler shifts, this indicates rapid turbulent motions with speeds up to a few thousand kilometres per second. An examination of the intensities of the spectral lines yields an estimate of electron density at about 10^7 electrons/cm^3 and a temperature around 30,000 K for 3 C 273, and comparable values for 3 C 48. The emission lines do not show evidence of rapid variability, and the size of the hot gas cloud is estimated to be several light years in diameter. We may thus begin to build up a model of the quasars with a nucleus less than a light year in diameter and yet of tremendous luminosity, surrounded by a very hot cloud of gas several light years in diameter.

Photographs of 3 C 273 taken with very long exposures with the 5 metre Mount Palomar telescope reveal a remarkable jet of nebulosity obviously having been ejected from the stellar object, as illustrated in the diagram, Fig. 9.18. The jet is visible up to 20″ away from the starlike nucleus, corresponding to a separation in distance of about 300 thousand light years.

Both the jet and the nucleus of 3 C 273 coincide with radio sources. The jet is the larger radio source, designated component A, and has a spectral index of 0.7, while the stellar object corresponds to a very small diameter source, component B having a quite different spectrum. These two components are illustrated in Fig. 9.18.

The radio source 3 C 273 therefore consists of two separate regions producing radio emission, one near the quasi-stellar object (B), the other a larger region near the jet (A). As the spectrum indicates, at short wavelengths of 10 cm or less, the concentrated region, 3 C 273 B, is predominant. In addition to the changes in optical brightness, radio variations also occur in 3 C 273 B and these are more marked at shorter

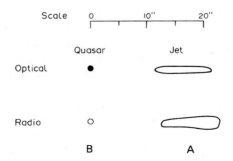

Fig. 9.18. Sketch of the optical and radio components of the quasar 3 C 273.

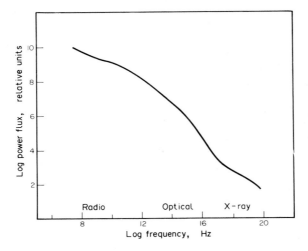

Fig. 9.19.   Spectrum of 3 C 273 from X-rays to radio.

wavelengths. The radio emission fluctuates less rapidly than the light, so the radio source may be assumed to be larger than the optical object. We shall discuss later the detailed structure and changes in the radio core.

3 C 273 has now been detected as an X-ray source, and Fig. 9.19 illustrates the spectrum in power flux per unit bandwidth over the whole range from X-rays to radio. The nature and continuity of the spectrum is strong evidence that the same radiation process applies to all the radiation bands. Taking into account the spectral bandwidths we find that the power radiated is similar in the X-ray, optical, and radio bands, giving a total luminosity of approximately 5×10^{39} watts.

3 C 48 differs from 3 C 273 in several respects. There is no optical jet, although there are extensive wisps of nebulosity with an extent of about 300 thousand light years. A single radio source coincides with the quasi-stellar object. Both optical quasar and radio source show variability. The energy radiated by 3 C 48 is less than 3 C 273 but is still in the same energy category as the strong radio galaxies. Again there is the additional mystery of extremely small size and intense radiation so characteristic of quasars. Recent observations show that the optical nebulosity mainly consists of stellar-type radiation apparently associated with a surrounding galaxy.

Structures of Quasars

The radio structures are generally difficult to decipher, partly because the radio sources are often inherently small, and partly because their distances

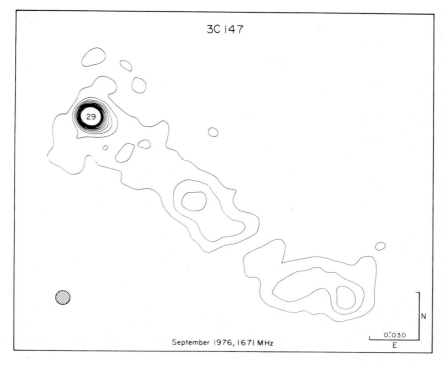

FIG. 9.20.   VLBI radio map at $\lambda = 18$ cm of the core and jet of 3 C 147. Contour levels 2, 6, 10, 14, etc., $\times 10^8$ K. Mean value in core 29×10^9 K. (After Readhead and Wilkinson, 1980.)

are so great. Consequently the angular diameters we wish to determine are usually very small, often less than 1″. Optical resolution is limited to about this figure by atmospheric turbulence causing "twinkling" so we do not know the true optical diameter by direct measurement.

Great ingenuity has been shown in attempts to determine radio structure. For quasars the bulk of the radiation arises from very compact regions, and the task of structural analysis is particularly acute. Various methods have been employed in the quest for data on sizes and structures. Inferences have been drawn from lunar occultations of sources, from intrinsic variability, from scintillations imposed by radio propagation through interstellar and interplanetary media, and from self-absorption of emitted radiation in the spectra. All these methods have made a valuable contribution to our knowledge. But in recent years the most exciting detailed information has been derived by very long baseline interferometry (VLBI) aided by analytical techniques such as "CLEAN" and "Closure Phase" as described in Chapter 3. In this way, maps with millisecond of arc resolution have been constructed.

The quasar 3 C 147, redshift $z = 0.545$, is one of the most intense sources at decimetre wavelengths, and has been progressively studied by various methods. The radio spectrum, Fig. 9.21, shows a sharp fall at longer wavelengths indicative of self-absorption of synchrotron radiation, which led Williams at Cambridge to predict a size of $\sim 0''.2$. Subsequently, long baseline interferometer measurements between Jodrell Bank and Malvern indicated a double structure with a separation $\sim 0''.2$. Meanwhile, interstellar scintillation observations at Cambridge had also implied angular dimensions of $0''.3 \pm 0''.1$. More recent interferometric measurements have established the morphology far more clearly.

Predominant at metre wavelengths are two outer lobes about $0''.7$ in extent. The central region, strongest at short wavelengths and successfully mapped by VLBI consists of a core and jet extending to about $0''.2$ as shown in Fig. 9.20.

The jet resembles that of the galaxy Virgo A except that the radio luminosity of the 3 C 147 jet is some 10,000 times greater. The relative spectral contributions of core, jet and outer lobes are shown in Fig. 9.21.

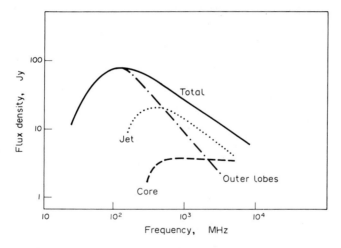

FIG. 9.21.   Radio spectra of components of the quasar 3 C 147. (After Wilkinson, Readhead, Purcell and Anderson, 1977)

The compact radio core of the quasar 3 C 273 has proved to be a most fascinating subject for study by means of VLBI, providing maps with millisecond angular resolution equivalent to linear dimensions of a few light years. The core structure is elongated toward the direction of the 3 C 273 jet. The core consists of a principal condensed component and minor components apparently moving outward with velocities greatly exceeding

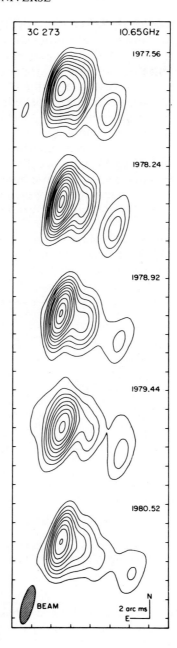

FIG. 9.22.   Apparent superluminal expansion of the core of the quasar 3 C 273 derived from VLBI data at λ = 2.8 cm. (After Pearson *et al.*, 1981.)

the velocity of light. The changes recorded by VLBI methods during the years 1977 to 1980 are shown in Fig. 9.22. The apparent rate of separation is about 10 times that of light. We shall consider later other examples of superluminal expansion, and the possible interpretation of these remarkable and disturbing results that seem at first sight to transgress a fundamental principle, that energy or matter can never move faster than light.

Another example of a quasar structure, 3 C 380, is illustrated in Fig. 9.23. The radio core mapped with the aid of VLBI, has been identified with an optical quasi-stellar object of redshift $z = 0.69$. The contours of the large-scale outer structure of 3 C 380 shown in Fig. 9.23 have been derived with the Cambridge 5 km radio telescope. A further example is the radio

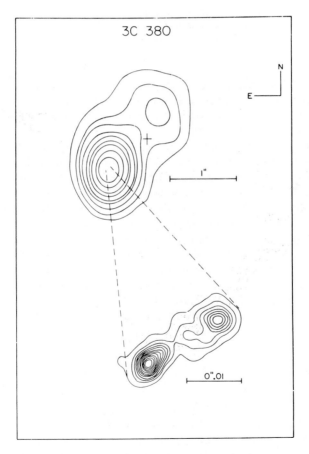

FIG. 9.23.   Radio maps of the components of quasar 3 C 380. The large-scale structure derived by Scott, 1977, at $\lambda = 2$ cm with the Cambridge 5 km telescope, and the core derived by VLBI at $\lambda = 18$ cm. (After Readhead and Wilkinson, 1980.)

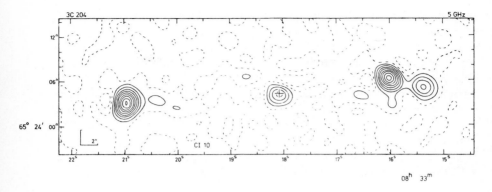

Fig. 9.24.   Radio structure of 3 C 204 mapped at λ = 6 cm with the Cambridge 5 km radio telescope. The central component coincides with the quasi-stellar object marked with a cross. (After Pooley and Henbest, 1974.)

map of 3 C 204 shown in Fig. 9.24, also obtained at Cambridge, which shows the symmetrical radio structure of a distant quasar, redshift $z = 1.112$. The disposition of components is clearly reminiscent of the prototype radio galaxy Cygnus A.

It is evident that the same types of structure associated with radio galaxies also occur in quasars. One exception appears to be the swept-back trail formation found so far only in certain radio galaxies within clusters. Statistically, of course, there are characteristic differences between quasars and radio galaxies. The radio luminosities of quasars are systematically greater. There is a tendency for more asymmetric structures in quasars, and about 10 per cent are extended on one side only. However, the most striking difference is that in general the radio cores of quasars are stronger, and a large proportion of quasars apparently have little or no extended structure. Relatively flat spectra are common in quasars, as might be expected for radiation predominantly associated with compact cores. Variability is also prevalent, indicative of small, intensely active sources at the nucleus.

BL Lacertae

In 1929 a variable "star" in the constellation Lacerta (the Lizard) was designated BL Lacertae. This early stellar clasification was brought into question in 1968 when the object was also found to be a radio source showing rapid variations of intensity and polarisation but no emission lines. Then followed the discovery of other similar sources, which became known as BL Lacertae-type objects, or for short, BL Lac objects. Although it was suspected that they might be very remote sources, the absence of spectral

lines initially made it impossible to establish their distances. Eventually during the 1970s, with the aid of image tube scanners, very weak emission and absorption spectra were detected in several BL Lac objects, so establishing that they are indeed very distant like the quasars. By 1980 about 60 BL Lac objects were known. Their outstanding properties are starlike optical appearance, spectral lines very weak or absent, and extreme variability of optical infrared and radio emission. Most of the radio flux comes from a compact core less than 1 arcsec in angular size, and usually about half the flux from a region of the order of a milliarc sec. About 50 per cent of the radio sources also possess large-scale structures. Both in their large- and small-scale components the BL Lac objects are comparable in linear size to the quasars. The cores have notably flat spectra (spectral index values concentrated near zero) whilst the extended components have steep radio spectra (index $\sim$ 1). Optical and radio luminosities are similar to those of quasars. In fact BL Lac objects are like violently variable quasars which lack sufficient surrounding gas to produce conspicuous spectral lines. Time scales of variability range from weeks to years, sometimes followed by long quiescent periods. Changes in radio power are typically by a factor of 2 but optical variations of about 100 times have been recorded. One BL Lac object flared up for several months in 1975 to attain the highest known luminosity of any object in the universe.

Apparent Superluminal Motion

One of the most provocative results emerging from very long baseline interferometry has been the evidence of apparent motion in central regions of extragalactic sources at velocities exceeding that of light. Often cores so far studied in detail by VLBI, four have shown superluminal speeds in the separation of components. These are the quasars 3 C 273, 3 C 279, 3 C 345 and the other a radio galaxy 3 C 120. The transference of energy at speeds exceeding that of light contravenes a fundamental tenet of modern physics and a basic principle of the theory of relativity. Considerable ingenuity has been shown in proposing explanations of the apparent motion. In the simplest model an expanding wavefront of radiant energy is incident on gaseous material and excites emission. An illusion of motion could ensue if the wavefront falls on different parts at successive instants. Such an event was recorded in 1901 when Nova Persei exploded, and a light echo traversed a nearby reflection nebula with a velocity of about 2 c.

In variable radio sources complications in interpretation could arise if changes of brightness occurred independently in different positions. However, the apparent superluminal expansion is well substantiated, with considerable consistency in rate and direction over periods of years. The observations of 3 C 273 have already been discussed on p. 210. Another example is 3 C 345; over the years 1970 to 1976 the radio core comprised

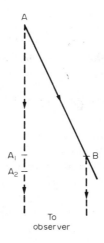

FIG. 9.25.   Observation of fast-moving source.

two similar components about 0.5 milliarcsec in size with their separation increasing from less than one to more than 2 milliarcsec over the 6 years. At the distance indicated by the redshift, $z = 0.59$, the speed of separation corresponds to nearly 7 c. After 1977 the western component weakened, giving the appearance of a diffuse elongated feature moving away from the compact eastern component (presumably associated with the quasar nucleus).

Examination of possible interpretations of these faster-than-light phenomena has drawn attention to the consequences of relativistic transformations. Theoretical study has demonstrated that sources moving with velocities close to that of light in directions within a narrow range of angles towards the observer can give the appearance of superluminal motion.

The following simplified illustration without involving the formulation of relativity may help in understanding how the illusion of faster-than-light movement is possible when a source approaches in a direction inclined to the line of sight. Figure 9.25 represents a source moving from A to B at a speed close to that of light. The observer interprets the projected displacement as a movement from A_1 to B. While the source travels to B the signal transmitted from A towards the observer reaches the position A_2. The interval between the observed signals received from A_2 and B is so brief that the calculated speed of apparent transverse motion (from A_1 to B) can exceed that of light. To be correct, of course, the implications of the theory of relativity must be incorporated. Nonetheless, the illustration demonstrates the idea as to how inferences of superluminal motion may arise.

The Quasi-stellar Objects

The generic term of quasi-stellar includes all very distant "starlike" objects, meaning that the major part of the optical image lies within a circle of 1″ diameter. The abbreviation QSO is used to classify all extragalactic objects of quasi-stellar appearance. The QSOs are therefore very compact sources of exceptionally high optical luminosity, usually 100 times or more that of a normal galaxy. The principal optical properties are generally found to be:

(a) starlike appearance and high luminosity,
(b) radiation mainly attributable to synchrotron radiation,
(c) spectrum usually showing blue and ultraviolet excess,
(d) broad spectral lines in emission, narrower ones in absorption, and large redshifts,
(e) many variable in brightness.

The emission lines arise from a gaseous ionised atmosphere around a brilliant nucleus. Absorption lines, frequently with slightly different redshifts (attributable to some relative motion), reveal the presence of surrounding or intervening gas. BL Lacertae objects are exceptionally variable QSOs with very weak or absent emission lines.

At very high redshifts some modification of the defining colour characteristics has to be accepted. The observed colour is dependent on the extent of the spectral shift, so blue excess is not an inevitable criterion. During 1973 two sources were found with redshifts exceeding $z = 3$. The first of these, OH 471, with $z = 3.4$, was identified with a quasi-stellar object of neutral colour. At the time of writing the most distant known quasar has a redshift of 3.78.

Colours are, of course, also influenced by the intrinsic properties of individual sources. The compact radio source 1413 + 135 has been identified as an extremely red QSO, and is one of the strongest known emitters of far infrared and millimetre radiation. Rapid variability and lack of emission lines indicate a BL Lac type. Its association with a galaxy has enabled its redshift $z = 0.26$ to be determined. In such a source we infer that a cut-off in energy distribution of the synchrotron emitting electrons is responsible for the unusually red spectrum.

The term quasar applies to QSOs that are strong radio sources. In 1965 the realisation by Sandage that many objects thought to be blue stars with no observable radio emission are actually distant QSOs came as a surprise. The existence of a large class of radio-quiet QSOs was subsequently confirmed by measurements of redshifts. In fact, it now appears that over 95 per cent of optically selected QSOs are radio-quiet.

It is not easy to derive the true numbers in space of different classes of objects from their observed apparent distribution. For instance, objects

that are rare, and sparsely distributed throughout space, may be detected in quite large numbers if they are powerful radiators. They seem plentiful because we can detect them to great distances. This is why so many quasars have been detected because they have exceptionally high optical and radio luminosity. Yet there are very great distances between them as compared with other galaxies. If we take the two of three hundred most easily detected radio sources that have been identified we find that about a third of them are quasars. Actually, the number density, that is, the number occurring in a given volume of space, is very low.

As a very rough guide it seems that quasars are at least 10 times (possibly 100 times) more numerous than BL Lac objects. The space density of radio-quiet QSOs is about 100 times that of quasars, whilst radio galaxies are over 1000 times more common than quasars. About 10 per cent of giant elliptical galaxies are radio galaxies. Finally, normal galaxies are many thousand times more numerous than all other types.

The Nature of Quasars and Radio Galaxies

One may well ask, what conclusions have we reached about the evolution of galaxies and quasars? Do the quasars represent a stage in the production or collapse of galaxies? Although the answers still elude us the evidence points to a relationship between them. It is interesting that observers with the 5 m telescope at Mt. Palomar have recently recognised several instances of a galaxy apparently enveloping a quasar. It is reasonable to asume that had these objects been viewed at greater distances the quasar images alone would have been distinguishable.

The cosmological interpretation of quasar redshifts, that is, that distances are correctly determined according to Hubble's law, has often been questioned.

Another way of producing a redshift is by gravitation. A photon radiated from a very massive but condensed source rises through a very strong gravitational field. Consequently, it loses energy, just as a cricket ball loses kinetic energy when thrown up from the ground. Any appreciable energy loss of a radiated photon due to the drag of gravity would appear as a redshift.[1] The idea of gravitational redshifts in quasi-stellar sources has been rejected in the past because it did not appear to be substantiated in other ways. Although controversy over the interpretation of redshifts erupts from time to time, the consensus of opinion still seems to point convincingly to the evaluation of cosmological distances by redshifts according to Hubble's law.

We now summarise some of the main conclusions that have been reached. The outstanding characteristic of many radio galaxies is the

[1] The quantum energy hf is less when the frequency f is reduced, that is shifted towards the red.

formation of double lobes. About 70 per cent of structures are double extending in opposite directions from the optical galaxy. The large components, vast reservoirs of energy in the form of relativistic particles and magnetic fields, show no perceptible changes in shape or power. In contrast with radio galaxies the dominant radio power in quasars is generally concentrated in the core. In fact, whereas extended components are the hallmark of radio galaxies they have been detected only in about 50 per cent of quasars. Nevertheless, where they are observed, their structures are similar and radio maps cannot distinguish between them with certainty. Compact cores are present in radio galaxies although weaker than in quasars.

Where jets are observed in radio sources they may be double or single. The appearance of one-sided jets could well be the result of relativistic beaming of radiation. Where jets occur they are remarkably in line with outer structures.

The excellent alignment so often found between cores and inner and outer radio structures, the detection of jets emanating from cores, and the concentrated emission and activity in the cores, provide overwhelming evidence that the large radio sources are produced by beams of energy expelled from the nucleus. Such continual transference of power also accounts for the replenishment of energy to zones of optical synchrotron radiation along jets and radio emission from distant hotspots. The double structures and the maintenance of alignment over millions of years show that power is being channelled outwards in opposite directions along a unique axis. We conclude that such a unique line can be provided only by the spin axis of a rotating nucleus.

The most challenging problem is to account for the release of such vast energy. If the nucleus is the source how can such enormous energy originate from a region of the order of a light year, or in some cases a light month or less in size? The power output of quasars can be as much as 10^{39} or 10^{40} watts in particles and fields in radio, optical and X-ray bands. The total energy stored in relativistic particles and fields in large radio lobes has been estimated to reach 10^{54} joules. This is equivalent to converting the mass of 10^{7} suns into energy.

It is now generally agreed that gravitational collapse, with ultimate contraction into a black hole, must account for the vast energy associated with a core of such extremely small dimensions. The mechanism of energy release must of course occur outside the black hole as accreted matter swirls into it. The conservation of angular momentum will develop enormously high speeds of rotation in an accretion disk of infalling material. Here collisions, viscosity, and interactions will release the kinetic energy of material eventually captured in the black hole. The conversion of mass into energy can attain an efficiency of over 40 per cent. A mass equal to a few suns devoured per year could realise the power output required for

strong radio galaxies and quasars. With durations of millions of years the total mass trapped inside the black hole may well exceed 10^8 suns. Although the detailed processes of energy conversion and release are open to conjecture, the spin axis appears to present natural directions of least resistance for outward flow. This generalised model accounts for the small nucleus, the release of vast power and the alignment of source structures.

Although it is tempting to suggest an evolutionary sequence, the relation between the different types of extragalactic objects, quasars, radio-quiet QSOs, BL Lacs, radio galaxies, and normal galaxies, remains a subject for speculation. We do not know, for instance, whether radio-quiet QSOs become radio sources during some stage of their development. It is also possible for radio emission to be so directional that it may elude detection. We can be certain, however, that the travel time of radiation from sources of large redshift is so great that they provide evidence of the state of the universe in the distant past at an earlier epoch in the evolution of the universe. In the next chapter we consider further the contribution of radio astronomical information to cosmology.

10. Cosmology

COSMOLOGY is a study of the behaviour and distribution of matter in the universe as a whole. It is concerned also with questions as to whether the universe is evolving, and whether it originated at some finite time in the past. An alternative might be that the universe has always appeared to have the same general properties, the so-called "steady state". We shall here consider the bearing of radio observations on these difficult problems. Before doing this, we will describe in a simple illustrative way some basic cosmological facts and concepts.

As radio waves and light travel with the same finite velocity, we observe distant objects in the condition in which they existed a very long time ago. If a galaxy is 100 million light years away, this means that the light and radio waves have taken 100 million years to reach us. Therefore we do not know the present state of the object, only what it was like 100 million years earlier.

Hubble's law of "redshifts" is an observational result of great importance in cosmology. According to this law, described on page 182, observed wavelengths are increased by a fraction proportional to the distance of the source. If the fractional increase of wavelength $\delta\lambda/\lambda$ is denoted by z, then in accordance with recent measurements, the distance D given by Hubble's law is

$$D = 20{,}000 \; z \text{ million light years.}$$

Now the normal interpretation of a shift in wavelength is that it is the Doppler shift due to motion. Hence the natural interpretation of the shift to longer wavelengths, (or "redshift" as it is called) is that the galaxies are receding from us. The universe is therefore expanding with the most distant objects moving away with the greatest speed. If the rate of expansion[1] has always been the same we may infer that the universe started 20,000 million years ago. This reasoning suggests a simple explanation of Hubble's law if we assume that the primeval matter initially possessed a great variety of speeds. Since the time of creation, matter with the highest speeds will have travelled furthest, so we would expect to find the most distant objects receding the fastest.

---

[1] The adopted estimate of the rate of expansion is about 50 km/sec per Mpc (Mpc = megaparsec = 3.26 million light years). The value is often known as Hubble's constant.

In thinking of models of the universe there are complicating factors. There is the question of the true geometry of space, and the implications of the theory of relativity. There may also be a changing rate of expansion.

It seems natural to think of space as infinite and that the ordinary rules of Euclid's geometry would apply. Before it was realised that the universe is expanding, it was pointed out by Halley that if stars are distributed throughout infinite space, then it is difficult to explain why the sky is dark at night. This difficulty does not arise in an expanding universe because the redshift represents a loss of energy[1] in the radiation from distant stars. An evolutionary universe with fewer or less bright stars at great distances would also resolve the difficulty.

The idea of non-Euclidean space geometry can be illustrated by considering a curved surface, for instance, a sphere like the surface of the Earth. If we draw a triangle from the North pole to two points on the equator, each angle can be 90°. You cannot draw a triangle on a flat surface with every angle 90°. A curved surface triangle is shown in Fig. 10.1.

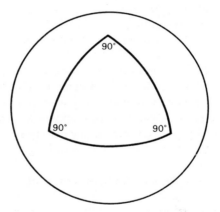

Fig. 10.1.   A triangle with 90° angles on a curved surface.

Imagine a spherical surface to represent a two-dimensional universe and that our observations are entirely limited to the surface. We could deduce by geometrical results obtained by surveying within the surface that we exist on a finite surface with no boundaries, and we could calculate its radius.

Einstein's theory of relativity showed that events in our actual world must be considered as specified by four dimensions, three being the spatial positions and the fourth the time. Distances and times are bound together by the observational fact that the velocity of light is always constant. The

[1] The conservation of energy is satisfied because work is involved in expansion.

resulting combination of space and time is called space-time. Einstein also showed that the motion of bodies under the influence of gravitation could be regarded as a consequence of the curvature of space-time produced by the presence of matter. Although we cannot visualise four-dimensional curved space-time, mathematical relations can be written down.[1]

Einstein found that his theory of gravitation could be used to construct a new static model of the universe only if he introduced into his equation a term corresponding to a force of cosmical repulsion that just balances the gravitational attraction of the universe. He found that in his model universe, three-dimensional space would be of finite extent, although it would have no boundary. He also found that the radius of this model would be proportional to the total mass. After the expansion of the universe was discovered, Einstein, in collaboration with de Sitter, realised that it was no longer necessary to introduce the idea of cosmical repulsion. Instead they used the theory of relativity to construct another model in which the expansion of the universe is gradually slowing down due to the gravitational attraction of the whole mass (just like a projectile shot from the Earth loses speed due to the Earth's gravitational pull).[2] Three-dimensional space is Euclidean in the Einstein–de Sitter model, but other models are possible in which the space is curved. In different models there are different relations between the mean density of the universe, the gravitational pull and the rate of expansion. For this reason experiments are often made to find the density of gas in intergalactic space in order to assess the mean density of matter in the universe. The simplest method is to try to detect absorption by the gas when observing very distant bright sources. The above discussion demonstrates that we are faced with a number of cosmological possibilities, and that the problem is complex.

In 1916 Einstein published a treatise '*The Foundation of the General Theory of Relativity*' formulating his new theory of gravitation in which space–time is curved near heavy masses. One suggested test of the theory is that it predicts a bending of 1".76 in a ray of light or any other form of electromagnetic radiation passing close to the Sun. The first experiment to test the bending of a light beam was performed in 1919 by the British astronomers Dyson and Eddington observing the positions of stars near the direction of the Sun during a total eclipse. The star positions were indeed shifted as predicted, and this verification of Einstein's theory aroused worldwide acclaim.

The accuracy obtained in the initial test was around 30 per cent. Modified theories of gravitation have subsequently been proposed and a

---

[1] The curvature enters into the equation through a factor k which has a value $+1$, 0, or -1 depending on whether the curvature is elliptical (or spherical), Euclidean, or hyperbolic (saddle shaped).

[2] The deceleration is usually represented by a symbol q, so defined that $q > \frac{1}{2}$ corresponds to a slowing down of the expansion.

much higher degree of accuracy is now required to distinguish between rival theories. More recent optical deflection experiments with an accuracy of 10 per cent are still insufficient.

In recent years radio methods have led the way in providing excellent confirmation of Einstein's formulation of the principles of General Relativity. To measure a deviation of 1″.76 at grazing incidence to the Sun is plainly within the scope of radio interferometry. The quasar 3 C 279 is occulted by the Sun in October each year and affords an excellent test source, while 3 C 273 about 9 ° away provides an appropriate directional reference. Several other sources have also been utilised. Allowance must, of course, be made for the influence of the solar corona on radio propagation. Coronal scattering widens the apparent diameter of a distant source and this limits the maximum baseline that can be employed. Bending due to coronal refraction can be reduced by choosing centimetric wavelengths, and corrections applied by observing at two wavelengths. Even so, solar activity as well as coronal effects set a limit to the closest permissible approach to the Sun for reliable measurements. Radio results have achieved high accuracy. For instance, observations at NRAO, USA, in 1974–5 with baselines up to 35 km, verified the gravitational bending predicted by Einstein to within about 1 per cent.

A slightly higher precision has been attained in another way from the delay caused by the solar gravitational field on radar signals travelling along paths close to the Sun. The delay can amount to 250 μsec for signal paths that graze the Sun. The first tests were made on planets at superior conjunction—that is, in close alignment on the far side of the Sun. Initially, the accuracy achieved on radar signals reflected from Venus and Mercury was about 5 per cent. More recently, the delays have been measured on signals relayed back from transponders on Viking spacecraft orbiting and landing on Mars. As in the deflection experiments compensation was made for the influence of the solar corona by the deployment of dual frequencies. The errors in determining the delays are only a few nanoseconds, hence the method has high potential accuracy. The results so far have confirmed the predictions of Einstein's theory to within 0.5 per cent.

One of the most sensational demonstrations of the gravitational bending of rays was the discovery of a double quasar consisting of almost identical twins whose appearance could only be ascribed to the existence of a gravitational lens somewhere along the line of sight. The possibility that doubles images of a very distant object might be formed by the space–time curvature of an intervening massive object had long ago been discussed in a paper by Einstein. In 1979 during spectroscopic studies of QSO objects identified with radio sources, it was noticed that a distant radio source, designated 0957 + 561, appeared to correspond to a pair of optical objects, 6″ apart, with the same redshift, and with such similar spectra that they are virtually indistinguishable. The coincidence was so exceptional that it could

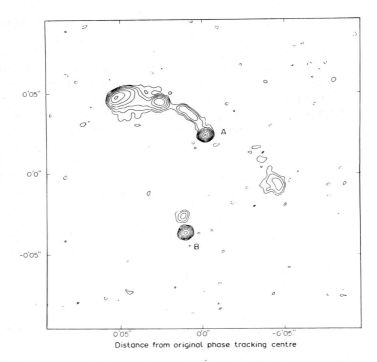

0'05"

0'0"

-0'05"

0'05"   0'0"   -0'05"

Distance from original phase tracking centre

FIG. 10.2.   VLA radio map at 6 cm wavelength of the double quasar 0957 + 561. (After Greenfield, Roberts and Burke, 1980.)

be inferred that in reality they were two images of the same object. It can be shown that if a distant source is viewed through the intervening gravitational lens of a massive yet compact object, exact alignment should result in a ring image; but in the more general case of offset alignment two images should appear.

When the radio structures were mapped in detail with the aid of long baseline interferometry, the two principal images designated A and B, were clearly evident as shown in Fig. 10.2. However, certain complexity in the vicinity of the northern image was puzzling. When later the lens-producing galaxy was optically recognised the problem of interpreting the inequalities was largely resolved. For the intervening object was discerned as an elliptical galaxy, redshift ≈ 0.4, centred in a direction close to B. It was then appreciated that the "lens" of this extended galaxy and its location could affect image formation so that complex structure appeared only near the northern image A. When the central regions of the images were examined by VLBI it was interesting to find that both possess precisely the same core-jet structure thus reinforcing their essential identity. The

formation of double quasar images with the same intensity in optical infrared and radio wavebands further establishes that the gravitational lens interpretation is correct. A year later another gravitational lens was discovered, on this occasion with three images hence called "the triple quasar". We may well expect that many more examples of gravitational image formation will be recognised.

Another impressive and intriguing demonstration of the validity of Einstein's theory followed from the analysis of data on the binary pulsar PSR 1913 + 16. A close watch on its behaviour has been maintained at Arecibo since its discovery in 1974. Neither the pulsar nor its partner is visible. The presence of the partner was revealed by systematic drifts in pulse recurrence rates attributable to Doppler shifts during a 7h 45m orbital motion. The pulse interval is 0.059 sec, one of the shortest of all known pulsars. The masses of the pulsar and its companion are each calculated to be approximately 1.4 solar masses, and it is concluded that both are neutron stars. In these unique circumstances the processes of atmospheric transfer are minimal.

The analysis of data fulfils two major predictions in accordance with Einstein's theory. The first relates to the precession of the orbit. The pulsar periastron, the position of closest approach to the companion, is found to precess at a rate of 4.2 degrees per year, a value in agreement with Einstein's theory. The second is a striking deduction that implies the radiation of gravitational waves. The binary pulsar incurs a very eccentric orbit and high velocities, and according to Einstein's theory a substantial amount of energy should be radiated as gravitational waves. In consequence the two bodies would be expected to draw closer together with a corresponding decrease in orbital period. The theoretically estimated decrease of orbital time due to gravitational radiation fits well the observed value of ~ 100 μsec per year. The results derived from the binary pulsar have been hailed both for their vindication of Einstein's theory, and for the first observational evidence of gravitational waves which hitherto had eluded attempts to detect them.

We will now turn to two particular aspects that have so far been given much attention by radio astronomers, namely number counts of radio sources, and measurements of background radiation from space.

A great deal of past controversy has been centred on the counts of sources in different ranges of received flux density ever since Ryle at Cambridge announced in 1955 that his results did not conform with expectations for a random distribution in a static Euclidean universe. If the radio source density is uniform throughout space, then calculation shows that the number of sources with received power influx density greater than S should be inversely proportional to $S^{1.5}$. This means that if we plot $\log N$ against $\log S$ we would expect a slope of magnitude 1.5. The early results gave a slope of 3, but this was later found to be erroneous, because of the

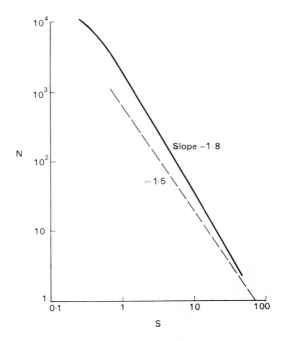

FIG. 10.3.   Number of sources, N, with power flux density greater than S; Cambridge observations, full line (slope -1.8). The dashed line (slope -1.5 for comparison would correspond to a uniform distribution). N is the number of sources per unit solid angle; S is the flux density at $\lambda = 1.7$ m in Jy.

effects of overlapping of sources within the beamwidth of the radio telescope. All such effects have now been carefully eliminated but the magnitude of the slope is still found to be greater than 1.5. The Cambridge results indicate a slope of 1.80 as shown in Fig. 10.3.

We note that the slope decreases for very small values of S. This is to be expected on almost any cosmological model that is not static Euclidean, because as we have already explained there must be a falling off in the contributions from very distant sources. We come then to the problem of accounting for the initial slope of 1.80, indicating an excess of sources of low flux density. Ryle has argued that this clearly suggests an evolutionary universe with numbers of radio sources greater in the past. The results could be interpreted as either greater numbers or greater luminosity of radio sources. In either case we would see more sources at greater distances, corresponding to earlier epochs, than we would expect on a uniform distribution. Analyses so far suggest that the excess in radio sources is largely confined to extended sources with steep spectra and high intrinsic luminosity.

In recent years radio observations of a different type have also brought strong support for an evolutionary cosmology. This is the observation of universal radiation. The background radiation we are familiar with in radio astronomy is the radio emission from the galaxy plus the general distribution of discrete radio sources. These contributions dominate the background at wavelengths longer than about 10 cm. Careful measurements at shorter wavelengths made by Penzias and Wilson in the USA in 1965 showed that there is an additional radiation appearing uniformly in all directions corresponding to a black body temperature of 3 K (3 degrees above absolute zero). In Fig. 10.4 the two research scientists are seen in front of the horn aerial used for their observations. This type of radio telescope minimises unwanted radiation from the ground.

FIG. 10.4.   A.A. Penzias and R.W. Wilson of Bell Laboratories, USA, in front of the horn aerial used in their discovery of cosmic background radiation.

The measurements have since been confirmed by other workers as illustrated in Fig. 10.5.

Dicke, Peebles and others have discussed the implications of these very important observations of universal radiation. They have shown that they fit well with a concept put forward by Gamow in 1946, that the universe

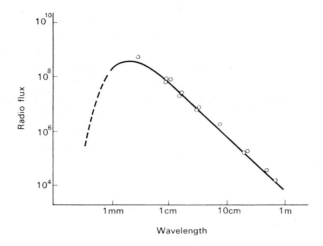

Fɪɢ. 10.5. The radio measurements (circles) of background radio emission fit well the radiation curve (full line) for a blackbody temperature of 27K. The Units of radio flux are 10^{-26} Wm^{-2} Hz^{-1} ster^{-1}. (After Shakeshaft and Webster, 1968.)

began in an intensely hot compressed state from which it has evolved and expanded. The observations fit this interpretation assuming an initial temperature of the order of 10^{10} deg K and the age of the universe about 10^{10} years. The redshift and cooling due to the expansion of the universe satisfactorily accounts for the low temperature of 3 K now observed. These radio astronomical results lend strong support to what is called the "hot big bang" theory of the beginning of the universe. There is one possible modification to the idea that this occurrence was the creation of the universe. Some cosmologists have suggested an oscillating model. According to this, the initial expansion would gradually slow down followed by contraction due to gravitational attraction, finally returning to a hot compressed state of the universe. Explosive conditions would then result in renewed expansion and so the process is repeated.

The Radio Universe

The preceding chapters have described the contribution of radio methods to astronomy. One may well ask, why has radio astronomy been so successful in making rapid and important advances in almost every field of astronomical science? In the first place, radio waves have the obvious advantage in their ability to penetrate through dust and haze that pervades interstellar space. The low energy of the radio quantum is another important factor. The quantum, the unit of energy interchange, given by $E = hf$, is small for radio waves because the frequency f is low compared with

that for light or X-rays. In consequence, radio has the merit that not only are radio photons (quanta) emitted by sources at low temperatures, but releases of energy in astrophysical phenomena easily result in the production of large numbers of low-energy photons and therefore of radio waves.

When we consider reception, ultimate sensitivity is set by the number of quanta received. Greater rates lead to lower percentage fluctuations. For a given bandwidth and power flux fewer quanta would be received at higher frequencies. However, certain practical factors have to be taken into account. At higher frequencies, such as those of X-rays, received bandwidths are greater and this increases the energy accepted. Also, X-ray detectors are very efficient photon counters. In comparison the ease by which radio waves are generated leads to an appreciable residual level of unwanted noise, partly external and partly within the receiver. When all compensating influences are balanced it transpires that distant active sources such as quasars may be detected to much the same depths in the universe by radio, optical radiation or X-rays. Indeed this is fortunate, for all wavebands have their special contribution to offer to the understanding of the physical processes involved. Nonetheless in the observation of many phenomena radio can claim a unique significance with outstanding achievements in sensitivity and resolution.

The great potentialities of radio as a means of communication, and the broadcasting of information by radio and television are well known to us. Now radio methods probe cosmic space bringing a vast amount of information about the universe. If there is life on other planets, and it has been estimated that some thousand million stars in the Galaxy may have planets in a condition to support some forms of life, then it seems possible that if there were any attempts at communication, these would also be made by radio waves (although the time delays of transmission constitute a major problem). By understanding and utilising the properties of radio waves and by discovering how to convert the waves into responses we can directly observe, radio has effectively provided man with an amazingly powerful new sense, a powerful means of communication, acquiring information, and exploring the universe.

What knowledge of the universe may we expect to derive? Shall we find the answer to the problem of the origin of the universe? Definitive answers to some cosmological questions will undoubtedly be found by further research, but experience from the history of science tells us that new problems continually arise. If, for example, the "big bang" theory is confirmed there is at once the question, how could it originate? The idea that science may provide an ultimate answer arises from a misconception. Reasoning brings a realisation of the limitations of scientific knowledge. We recall our discussion in Chapter 1 on the descriptive nature of science. The pursuit of astronomy gives a clearer descriptive picture of the physical framework of the universe we observe. In this way we may better

understand our relation to the universe. Man is only part of the universe and cannot be expected to comprehend the whole. If he could, he would be superior to the whole universe. All he can achieve by scientific research is a greater knowledge to help him to understand and fulfil his part. Whatever our cosmological view, the mystery, and purpose are beyond the realms of physical science.

Appendix A   Scales of Measurement

IN ASTRONOMY we observe sources at immense distances. The Sun is 93 million miles away, and the nearest star over 20 million million miles, and yet such distances are very close compared with many stars and galaxies that can be observed by optical or by radio telescopes. The fact that we can detect objects at such great distances means that they must be radiating tremendous energy. We are familiar with a 100 W electric lamp bulb, or 1 kW (1000 W) electric heater, but the power of an astronomical source must be many millions of millions of watts to be detected at the Earth. It is therefore essential in astronomy, both for distances and energy to have easy ways of expressing very large numbers. Suppose we have a large number such as 5 million million, that is, 5,000,000,000,000, then we write this as 5×10^{12}, meaning 5 multiplied twelve times by 10. Even if we find it hard to visualise what a very large number really represents, at any rate we have a convenient way of writing it down. For instance, the light power emitted by the bright star Sirius is about 6×10^{27} W (or 6×10^{24} kW).

Total energy over a given time can be expressed in kilowatt-hours, the familiar domestic unit for reckoning our electricity consumption. An alternative unit often used in physics is the erg (the Greek name for work). If we note that 1 kW is 10^{10} ergs per second we can easily convert our units as required.

Distances can be expressed in miles, or kilometres if we use the metric system (5 miles is practically equal to 8 km), but astronomical distances are so great that we normally employ larger units. One commonly used is the light year, namely the distance light travels in a year. As we know that the speed of light is 186,000 miles per second, we find that 1 light year = 6×10^{12} miles approximately, or 6 million million miles. In the metric system, the speed of light equals 300,000 km/sec and 1 light year = 9.5×10^{12} km. Although it may seem difficult to think what such large values represent, we can illustrate large numbers by familiar examples. For instance, the distance from Manchester to Birmingham is 120 km, but stated in cm it is 12×10^{6}, that is, 12 million cm.

The nearest star to us is 4.3 light years away and the brightest star Sirius is 8.8 light years. The stars that we see around us in the sky at night belong to our Galaxy, a system of stars of which our Sun is one member. It has been estimated that there are 10^{11} stars in our Galaxy. About 3200 of these are near enough to be distinguished on clear nights without a telescope in our hemisphere.

Curiously enough, when we come to measure the power we receive from distant astronomical sources we have the problem of expressing fantastically small quantities. As we explain later, the power we detect from a radio astronomical source might be only $1/10^{20}$ part of a watt. We write this as 10^{-20} W. Similarly, a millionth part could be written as 10^{-6}. So we see that the same method of notation is just as helpful in writing down very small numbers as it is for very large numbers.

Positions in the Sky

Directions in the sky are reckoned in much the same way as geographical locations on the Earth are denoted by means of latitude and longitude. Geographical lines of longitude run from North to South Poles, and $0°$ longitude is chosen as the line passing through London (Greenwich). The lines of latitude run parallel to the equator which is taken as $0°$ latitude, and the latitude of London is $52°$N.

In ancient times people thought the stars were points of light on a huge dome. Although we know that we are looking out into the depth of space, we still find it convenient when indicating directions to talk of the sky as the celestial sphere. Positions on the celestial sphere are given by their Declination (similar to geographical latitude), and by their Right Ascension (like geographical longitude). In terrestrial longitude we reckon from Greenwich as the starting point, calling it $0°$ longitude. In a similar way for Right Ascension, a zero position has to be chosen, and this point in the sky is where the Ecliptic (the apparent path of the Sun in the sky) crosses the

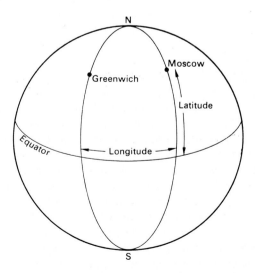

Fig. A.1.   Geographical positions measured in longitude and latitude.

celestial equator at the vernal equinox. It was originally called the First Point of Aries (denoted by symbol ♈, the ram's horns in the sign of the Zodiac).

The Earth spins about its North–South axis once per day, but of course we think in terms of ourselves and imagine that we are stationary and that the Celestial Sphere spins around us. Since a complete revolution of 360° is made in 24 hours, a rate of 15° per hour, it is usual to reckon Right Ascension hours instead of degrees.

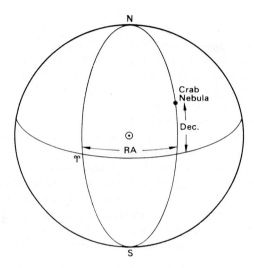

FIG. A.2.   Astronomical positions measured in Right Ascension (R.A.) and Declination (Dec.).

Solar Time and Star Time

Due to the Earth's rotation, the apparent position of a star with respect to ourselves depends on the time. There is an interesting difference between solar time, that we use everyday, and star time. It was in fact this difference that led Jansky to the conclusion that he had detected radio waves from the Galaxy. Solar time is measured from the passing of the Sun due South, namely 12 o'clock noon. The star positions at the same time appear to be shifted slightly each day. The explanation is as follows. The Earth moves in an orbit round the Sun once per year, and the diagram (Fig. A.3) illustrates the Earth travelling in its orbit from *A* to *B*, and also spinning about its axis. When the Earth has spun round exactly once, the observer at *O* faces a star in the same direction in the sky. This is a day as measured by the stars, and is called a sidereal day or 24 hours of sidereal time. (The word sidereal is derived from the latin "sidereus", meaning star-like.)

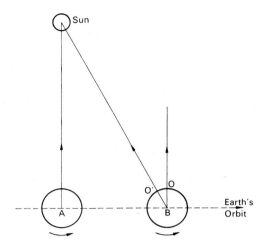

FIG. A.3.   Difference between sidereal and solar days.

The observer at O has to wait until the Earth's rotation brings him to O' before he faces the Sun again because the Earth has progressed in its orbit from A to B. Using solar time, as we do, the star positions at a given time, say 10 p.m., appear to drift by an amount equivalent to about 4 min per day.

Magnitudes and Luminosity

Another important quantity, is the power radiated by the source. In optical astronomy, stars are classified in apparent magnitude according to the light we receive from them. This practice dates from ancient times when Hipparchus arbitrarily graded stars in six magnitudes, and like places in a class list he numbered the brightest as first magnitude, and the weakest just visible to the unaided eye as sixth magnitude. Sir John Herschel in 1830 specified the magnitude scale more precisely, so that a change of 5 magnitudes represents a difference of 100 times in the amount of light received. With large telescopes, stars down to magnitude 19 can be observed visually, and with the aid of long exposure photographs much fainter stars can be recorded.

Knowing the apparent magnitude of a star and its distance, the luminosity, or total power of the light emitted by the star can be deduced. The calculation depends on the fact that the intensity of light diminishes inversely as the square of the distance from the source, a law which follows simply because the radiation spreads over a bigger area the farther it travels. Similarly, in radio astronomy, by measuring received power and

knowing the collecting area of our radio telescope and the distance of the source, we can calculate the radio luminosity, that is, the radio power produced by the source.

Suppose that with our radio telescope we determine the received power in watts per sq. metre. This is known as the power flux density received from the source. A typical value might be 5×10^{-26} W/m^2. Let S represent the power flux density and R the distance of the source. Assuming the source to be radiating in all directions, the power eventually passes outwards at distance R across a spherical surface of area $4\pi R^2$. Hence the total radio power P, emitted by the source is given by $P = 4\pi R^2 S$.

Appendix B   The Radio Receiver

FIGURE B.1 shows the essential stages of a system for measuring the power flux received from astronomical sources.

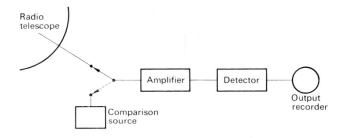

FIG. B.1.   Receiver system.

The radiation collected by the radio telescope is brought to a focus and fed by cables or waveguides (often called the feeders) to the amplifier. As the radio frequency signal oscillates rapidly it has to be converted by means of a detector into a direct current in order to produce an output that can be recorded. The received power can be measured by comparison with a reference source which can be connected to the input as required.

As the received signal is very weak it is passed through an amplifier.[1] We are anxious not to add appreciable noise from the equipment itself, and as any unwanted noise generated at the first stage undergoes the full amplification of the receiver it is important to have a low-noise first stage, or "pre-amplifier". Several kinds of low-noise amplifier are available; one type is known as the parametric amplifier, and another is the maser. Such pre-amplifiers often have to be cooled, so that they will work efficiently without producing thermal noise, and the best performance is obtained with a maser cooled with liquid helium.

---

[1] Usually the main amplifier incorporates a "frequency changer". The reason is that too much amplification at one frequency tends to produce instability and oscillation. This difficulty is overcome by amplifying not more than perhaps 1000 times at the radio frequency, and then changing to a lower frequency and continuing with further amplification at this lower frequency, called the "intermediate" frequency.

When a radio amplifier is tuned to a particular frequency, for example, 100 MHz, it also lets through a small range of frequencies on either side, say from 99 to 101 MHz. The frequency bandwidth in this case is therefore 2 MHz. If the amplifier is more sharply tuned, the bandwidth is narrower. In sound radio sets we use narrow bandwidths to keep out interfering stations, although the bandwidth must be sufficient to cover the range of frequencies present in a sound wave. In radio astronomy, however, we generally use wide bandwidths for the radio frequency amplifier, because, as we shall explain shortly, the receiver is then more sensitive. Needless to say interference can be troublesome. Radio astronomy sites are chosen in country areas where there is minimum interference. The situation is helped also by the protection of certain radio bands reserved for radio astronomy.

The action of the detector which follows the amplifier is best understood by considering its purpose in the ordinary sound radio receiver. Suppose a radio set is receiving a radio wave which is modulated in amplitude at 1000 c/s. If we wish to convert the modulation into a 1000 c/s vibration so that we can hear it, then we must use a detector. Figure B.2 illustrates the function of the detector in three stages:

(a) The incoming wave on the aerial produces alternating voltage modulated in strength 1000 times per second.

(b) The detector allows current to flow during positive half cycles of wave.

(c) The electrical pulses flow through the loudspeaker which responds only to the mean current giving an audible note at 1000 c/s.

If we send out two radio frequencies of steady strength separated by 1000 c/s, the two radio waves, being at slightly different frequencies would

(a)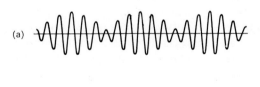

(b)

(c)

Fig. B.2.   Detection of modulated signal.

go in and out of step 1000 times a second. They are said to "beat" together, and if they were received on a radio set we would again hear 1000 c/s note. We see then that the radio signal with 1000 c/s modulation in Fig. B.2(a) is equivalent to two continuous waves separated by 1000 c/s.

Now the radio waves radiated by astronomical sources are occurring at many different frequencies at random because they are being generated by vast numbers of electrons in the source moving in all directions. The radio frequencies consequently "beat" together not simply at one note but a multitude of irregular notes, so by listening to the detected radiation one hears a random noise. The term radio noise is commonly used to describe such random signals and the low hissing noise they produce.

We can deduce how fast these irregular "beats" occur. We have seen above that the two radio frequencies 1000 c/s apart beat together 1000 times per second. Now the radio frequencies received together in a radio astronomical receiver depend on the bandwidth of the receiver. If the bandwidth is say 2 MHz then all frequency differences up to 2 MHz are received, and therefore irregular variations up to 2 MHz, that is 2,000,000 times a second, occur in the detected signal.

In radio astronomy we do not normally listen to the received signals. The output from the detector can be measured on a voltmeter of the pen-recording type, or often nowadays the readings appear on punched paper tape. The output recorder is arranged to have a slow response, so that it averages the output over an interval known as the time constant, which is usually about a second or more.

A radio astronomy receiver is made sensitive by having a wide input bandwidth for the radio frequencies, and a long output time constant. We will now proceed to explain why this gives more sensitivity. The wide input bandwidth results in a high rate of irregular beats. As we have shown above, if the bandwidth is 1000 c/s then there are about 1000 irregular beats per second. The detector converts these radio beats into one-directional current pulses, and the output recorder measures their average strength over the response time of the recorder. If the output response time is 10 sec, and there are 1000 pulses per second from the detector, then the mean amplitude of 10,000 pulses is recorded. The more pulses there are, the steadier is this average, so making it easier to detect any change of the mean level. For a sensitive receiver we therefore have a wide input bandwidth to let in a large number of pulses per second, and a slow output so as to average over a long time.

Suppose we are measuring radiation corresponding to a temperature of 27°C, or 300° above absolute zero. (Some of this will be received in the radio telescope and some will be generated in the receiver itself.) If the receiver bandwidth is 1 MHz, and the output time constant is 1 sec, then each significant reading of the output averages a million pulses. Statistics tell us that the output variation will only be about 1 part in $\sqrt{1,000,000}$ that

is 1 part in 1000, from the mean level. Hence our measurement of 300 degrees temperature will be accurate to

$$\frac{1}{1000} \times 300° = 0.3°.$$

A really sensitive radio astronomy receiver may easily have sufficient sensitivity to be able to detect 0.01 deg. If we wish to be as sensitive as this, it is best to keep comparing the received signal with our standard continually during the measurement, so we shall not then be misled by drifts in the amplification of the receiver. This switching technique is illustrated in Fig. B.3.

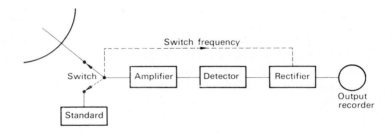

FIG. B.3.   Switching radiometer.

The rate of switching to the standard can be rapid, say 50 times a second. Of course the signal as it emerges from the detector will then be varying at 50 c/s. We can easily convert this into a steady output by means of a rectifier as indicated in the above diagram.

Low-noise Receivers

To attain very high reception sensitivity it is essential to minimise the noise generated within the receiver. The invention of the maser amplifier in the 1950s led to a marked improvement. The principle of maser action has been described in Chapter 2. A "pump" oscillator can produce over-populated energy levels in chosen maser material, ruby for example. An incoming signal can then induce stimulated emission with resulting amplification. A practical receiving device cooled in liquid helium can be designed to provide amplification of several thousand times while contributing to noise equivalent to no more than 10 degrees above absolute zero.

An alternative rather simpler although less efficient method of low-noise amplification is provided by the parametric amplifier. Essentially energy is

extracted from an injected "pump" oscillator by means of a non-linear reactance in order to amplify the incoming signal. The possibility of this method was predicted as early as 1931 by Lord Rayleigh, but not until 1957 was a practical device successfully constructed.

Radio Spectrometers

The simplest way of measuring a radio spectrum is by tuning a narrow band receiver successively to different frequencies. If a signal varies, as in a solar burst, a rapid sweep in frequency is required. For a steady signal, as in the 21 cm H line, greater sensitivity is achieved by observing over longer periods. However, such a method is comparatively slow and laborious. The next advance was the development of the multi-channel spectrometer in which a bank of filters divides the whole received bandwidth into a series of narrow bands. Long-term stability and calibration present problems.

A different approach, now more widely used, has been facilitated by progress in computing techniques. All the essential information in a received signal is supplied by sampling it at a rate of twice the bandwidth. If this is 5 MHz, the sampling rate is 10 MHz, well within the scope of modern computers. Suppose there is a high intensity at a certain frequency; it follows that the corresponding periodicity is strongly represented within the apparently random signal. The periodicity can be discerned by finding the correlation of signal intensities at given interval steps. The autocorrelation function can be derived for different intervals by feeding the signal into a series of delay channels followed by a computer. In this manner the intensities for different periodicites can be deduced. The method is equivalent to a Fourier analysis of the signal. The spectrum is in fact the Fourier transform of the autocorrelation function. The technique has valuable advantages for rapid and convenient spectral analysis.

Useful Data

VALUES are given to a convenient approximation.

Multiplying Factors

	Name	Symbol		Name	Symbol
10^9	giga	G	10^{-9}	nano	n
10^6	mega	M	10^{-6}	micro	μ
10^3	kilo	k	10^{-3}	milli	m
10^2	hecto	h	10^{-2}	centi	c
10	deca	da	10^{-1}	deci	d

Length

1 light year (ly) = 9.45×10^{12} km.
1 Astronomical Unit (A.U.) = 149,597,870 km $\approx 1.5 \times 10^8$ km.
1 parsec (pc) = 3.26 light years $\approx 3 \times 10^{13}$ km $\approx 2 \times 10^5$ A.U.

Angular measure

1 radian = 57.3 degrees (°).
1 degree = 60 minutes of arc (′).
1 minute of arc = 60 seconds of arc (″).

Velocity of Light (c)

c = 300,000 km/sec = 3×10^8 m/sec.

Energy and Power

1 joule (J) = 10^7 ergs.
1 watt (W) = 1 joule/sec.
1 electron volt (eV) = 1.6×10^{-19} joules.

Power Flux Density

This is the power (watts) flowing across 1 sq.metre (m^2) per cycle/sec (Hz) of bandwidth.
Hence the units are $Wm^{-2} Hz^{-1}$.
A more convenient unit in radio astronomy is the flux unit called the jansky (Jy) equal to $10^{-26} Wm^{-2} Hz^{-1}$.

Electron

Mass = 9.1×10^{-28} g.
Charge (e) = -1.6×10^{-19} coulombs.

Planck's constant (h)
   $h = 6.6 \times 10^{-34}$ joule sec.
Boltzmann's constant (k)
   $k = 1.38 \times 10^{-23}$ joule/deg K.
Magnetic Flux Density (T)
   1 tesla (T) $= 10^4$ gauss.
Area
   1 are (a) $= 100$ m^2 1 hectare (ha) $= 10,000$ m^2 ($= 2.47$ acres).
Mass
   1 tonne $= 1000$ kilograms ($= 0.984$ tons).

Acknowledgements

IN THE preparation of the new edition I am especially indebted to J.R. Shakeshaft for detailed comments and advice on extragalactic sources, to R.D. Davies for suggestions on galactic radiation, to P.G. Murdin for information on radio stars, and to L.T. Little for a discussion on the influence of scattering on radio observations. I also wish to thank those who have sent useful reprints and papers, particularly H. van der Laan, G.K. Miley, R.G. Strom, A.C.S. Readhead, S.H. Zisk, G.H. Pettengill, G.A. Giles, D.J. Helfand and R.C. Jennison. The reprints provided by the National Radio Astronomy Observatory, USA, and Jodrell Bank, Manchester University, have also proved extremely helpful sources of scientific papers.

In addition to the acknowledgements made in previous editions I now wish to express my gratitude to all who have supplied photographs, diagrams, and information for this third edition. Many of the illustrations are free adaptations from Figures in scientific journals and books, and appropriate references to the authors with dates are included in the captions. I am grateful to the editors and publishers of these scientific publications for permitting reproduction of the illustrations. The sources of new or revised Figures in this edition are as follows:

Figs. 8.5(b), 9.12, 9.13, 9.16, 9.21, 9.22 from *Nature* (Macmillan Journals Ltd.).

Fig. 9.11 from *Physica Scripta*.

Fig. 8.14 from the *Astronomical Journal* (American Astronomical Society).

Figs. 9.20, 9.23 from *The Astrophysical Journal* (American Astronomical Society).

Figs. 6.5, 10.2 from *Science* (American Association for the Advancement of Science).

Figs. 8.6, 8.7, 8.9(b), 9.2, 9.4, 9.15, 9.17 from *Astronomy and Astrophysics*.

Figs. 5.4, 5.9, 5.18 from *Solar Physics* (Reidel Publishing Company).

Fig. 5.17 from *Highlights of Astronomy* 1974 (I.A.U.).

Fig. 9.10(b) from *Ann. Rev. Astron. Astrophys.* 1980 (Annual Reviews Inc., USA).

Fig. 9.9 from *Proc. IREE* (Institution of Radio and Electronic Engineers, Australia).

Fig. 5.13 from *Proceedings of the Astronomical Society of Australia*.

Figs. 8.3(b), 8.5(a), 9.13, 8.8, 9.24 from Monthly Notices of the Royal
Astronomical Society.

Copies of photographs and diagrams have been supplied by courtesy of
the following, and I am most grateful for their kind and generous
cooperation.

Figs. 7.1, 8.1, 8.3, 8.9(a), 9.9(a) (Hale Observatory photograph) Royal
Astronomical Society.
Fig. 9.10(a) (Lick Observatory photograph) Royal Astronomical Society.
Fig. 5.3 Royal Greenwich Observatory.
Fig. 5.2 (CSIRO Solar Observatory photograph) R.J. Bray, D. Johns and
W. Place.
Figs. 1.1, 10.4 (Bell Laboratories, USA) C.F. Morgan.
Fig. 10.2 (Bell Laboratories, USA) P.E. Greenfield.
Fig. 3.11 (Arecibo Observatory) G.A. Giles.
Figs. 3.7(f), 6.5 (Haystack Observatory, USA) S.H. Zisk.
Fig. 8.5(b) (Columbia Univ., USA) D.J. Helfand.
Fig. 3.7(c) (Manchester Univ.) A.C.B. Lovell.
Figs. 3.24(a), 8.5(a), 8.8, 9.3 (Cambridge) J.R. Shakeshaft.
Fig. 9.13, 9.20, 9.23 (Cal. Tech., USA) A.C.S. Readhead.
Fig. 9.22 (Cal. Tech., USA) T.J. Pearson.
Fig. 9.9(d) (Sydney) W. N. Christiansen and D.J. Skellern.
Fig. 3.26 (Sydney) B.Y. Mills.
Figs. 5.4, 5.5, 5.9, 5.13, 5.17, 5.18 (CSIRO) R.X. McGee.
Fig. 3.28 (CSIRO) H.C. Minnett.
Figs. 3.7(a), 3.25, 8.7, 8.14, 9.10(b), 9.11, 9.12, 9.15 (National Radio
Astronomy Observatory, operated by Associated Universities Inc.,
under contract with the National Science Foundation USA) R.J.
Havlen and W.B. Weems.
Fig. 8.6 (Dwingeloo) R.G. Strom.
Fig. 3.24(b) (Westerbork) A.G. Willis.
Fig. 9.17 (Leiden) G.K. Miley.
Fig. 9.2 (Kepteyn Laboratorium) R. Sancisi.
Fig. 9.4 (Kapteyn Laboratorium) P.C. van der Kruit.
Fig. 8.9(b) (Leiden) H. van der Laan.

(I am also indebted to G.K. Miley for his help in obtaining Figures from
the Netherlands.)
I greatly appreciate the assistance received at the Royal Greenwich
Observatory in obtaining and preparing photographs and Figures, and I am
especially grateful to D.A. Calvert, J. Dudley and M.J. Everest. I am also
indebted to the library staff for their excellent service. I wish to thank the
Director of the Royal Greenwich Observatory for the help and facilities
provided at Herstmonceux.

Index